T0136994

Data Analytics

Series editors
Longbing Cao, Advanced Analytics Institute, University of Technology, Sydney,
Broadway, NSW, Australia
Philip S. Yu, University of Illinois at Chicago, Chicago, IL, USA

Aims and Goals:

Building and promoting the field of data science and analytics in terms of publishing work on theoretical foundations, algorithms and models, evaluation and experiments, applications and systems, case studies, and applied analytics in specific domains or on specific issues.

Specific Topics:

This series encourages proposals on cutting-edge science, technology and best practices in the following topics (but not limited to):
Data analytics, data science, knowledge discovery, machine learning, big data, statistical and mathematical methods for data and applied analytics,
New scientific findings and progress ranging from data capture, creation, storage, search, sharing, analysis, and visualization,
Integration methods, best practices and typical examples across heterogeneous, interdependent complex resources and modals for real-time decision-making, collaboration, and value creation.

More information about this series at http://www.springer.com/series/15063

Iván Palomares Carrascosa
Harsha Kumara Kalutarage • Yan Huang
Editors

Data Analytics and Decision Support for Cybersecurity

Trends, Methodologies and Applications

 Springer

Editors
Iván Palomares Carrascosa
University of Bristol
Bristol, UK

Harsha Kumara Kalutarage
Centre for Secure Information Technologies
Queen's University of Belfast
Belfast, UK

Yan Huang
Queen's University Belfast
Belfast, UK

ISSN 2520-1859
Data Analytics
ISBN 978-3-319-86624-6
DOI 10.1007/978-3-319-59439-2

ISSN 2520-1867   (electronic)

ISBN 978-3-319-59439-2   (eBook)

Printed on acid-free paper

This Springer imprint is published by Springer Nature
The registered company is Springer International Publishing AG
The registered company address is: Gewerbestrasse 11, 6330 Cham, Switzerland

Preface

Cybersecurity has been classically understood as the protection of any form of information potentially exposed in the Internet. This notion of cybersecurity has progressively evolved towards an umbrella term, covering a broad range of areas concerning the protection of hardware, software, networks and their underlying information and considering the socio-economic, legal and ethical impact that failing to guarantee such protection may cause on systems, data and users. A vital aspect of cybersecurity is to ensure that *confidentiality*, *integrity* and *availability* requirements are met for assets and their related users.

Both under a developmental and research perspective, cybersecurity has undoubtedly attained an enormous importance within the last few years. This is largely a consequence of the massive growth experienced by available data in cyber space, in their volume, value, heterogeneity and, importantly, degree of exposure. This explosion of accessible data with diverse nature accentuates the potential—and sometimes critical—vulnerability of the information harboured by them. Specifically, such vulnerabilities become critical when valuable information or knowledge might be illegitimately accessed by the wrong person/s (e.g. attackers) with malicious purposes. Unsurprisingly, both researchers and practitioners presently show an increasing interest in defining, developing and deploying computational and artificial intelligence (AI)-based approaches as the driving force towards a more cyber-resilient society.

With the advent of big data paradigms in the last few years, there has been a rise in data science approaches, raising in parallel a higher demand for effective data-driven models that support decision-making at a strategic level. This motivates the need for defining novel data analytics and decision support approaches in a myriad of real-life scenarios and problems, with cybersecurity-related domains being no exception. For instance, Fig. 1 illustrates the inter relationship between several data management, analytics and decision support techniques and methods commonly adopted in cybersecurity-oriented frameworks in the last years.

This edited volume comprises nine chapters, covering a compilation of recent advances in cybersecurity-related applications of data analytics and decision support approaches. Besides theoretical studies and overviews of existing relevant literature,

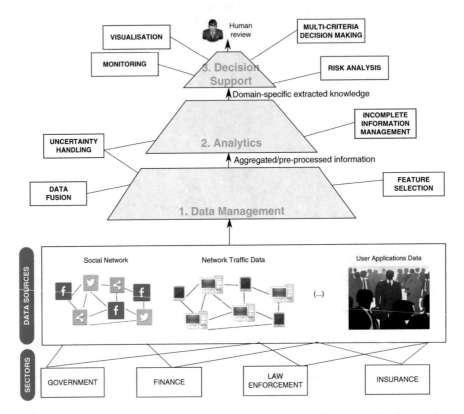

Fig. 1 Overview of data analytics and decision support processes and techniques in cybersecurity scenarios

this book comprises a number of highly application-oriented research chapters. The investigations undertaken across these chapters focus on diverse cybersecurity problems and scenarios situated at forefront directions of interest by academics and professionals. The ultimate purpose of this volume is twofold:

1. To bring together and disseminate some important "lessons learnt" within a young but surprisingly rapid field
2. To emphasize the increasing importance that data science and AI techniques (particularly those related to machine learning, data visualization and decision support) are attaining to overcome the major limitations and challenges currently arising in the IT security landscape

Therefore, this book joins ideas and discussions from leading experts in these rapidly growing fields, in order to align key directions of work that researchers and professionals alike would be encouraged to follow to further consolidate such fields.

Part I—Regular Chapters The first seven chapters present both theoretical and practical-industrial contributions related to emergent cybersecurity research. Particular emphasis is put on data analysis approaches and their relationship with decision-making and visualization techniques to provide reliable decision support tools.

In Chap. 1 [1], Markus Ring et al. present a novel toolset for anomaly-based network intrusion detection. Motivated by challenges that frequently hinder the applicability of anomaly-based intrusion detection systems in real-world settings, the authors propose a flexible framework comprising of diverse data mining algorithms. Their approach applies online analysis upon flow-based data describing meta-information about network communications, along with domain knowledge extraction, to augment the value of network information before analysing it under multiple perspectives. To overcome the problem of data availability, the framework is also conceived to emulate realistic user activity—with a particular focus on the insider threat problem—and generate readily available flow-based data.

Legg reflects in Chap. 2 [2] on the problem of detecting insider threats in organizations and major challenges faced by insider threat detection systems, such as the difficulty to reduce false alarms. The chapter investigates the importance of combining visual analytics approaches with machine learning methods to enhance an iterative process of mutual feedback between detection system and human analyst, so as to rationally capture the dynamic—and continuously evolving—boundaries between normal and insider behaviours and make informed decisions. In this work, the author demonstrates how generated visual knowledge can significantly help analysts to reason and make optimal decisions. The chapter concludes with a discussion that aligns current challenges and future directions of research on the insider threat problem.

Malware detection is no longer a problem pertaining to desktop computer systems solely. With the enormous rise of mobile device technologies, the presence and impact of malware have rapidly expanded to the mobile computing panorama. Măriuca et al. in Chap. 3 [3] report on Android mobile systems as a potentially major target for collusion attacks, i.e. attacks that result by "combining" permissions from multiple apps to pave the way for attackers to undertake serious threats. The authors present two analysis methods to assess the potential danger of apps that may become part of a collusion attack. Apps assessed as suspicious are subsequently analysed in further detail to confirm whether an actual collusion exists. Previous work by the authors is adopted as a guideline to provide a general overview of the state-of-the-art research in app collusion analysis.

In Chap. 4 [4], Carlin et al. focus on a recent strategy to fight against the devastating effects of malware: the dynamic analysis of run-time opcodes. An opcode is a low-level and human-readable machine language instruction, and it can be obtained by disassembling the software program being analyzed. Carlin et al. demonstrate in their work the benefits of dynamic opcode analysis to detect malicious software with a significant accuracy in real practical applications. One of such notable advantages is the ability of dynamic analysis techniques to observe malware behaviour at runtime. The model presented by the authors, which uses n-gram analysis on extracted opcodes, is validated through a large data set containing highly representative malware instances. Results show a superior malware classification accuracy, when compared with previous similar research.

Moustafa et al. in Chap. 5 [5] present a scalable and lightweight intrusion detection system framework, characterized by using a statistical decision-making engine to identify suspicious patterns of activity in network systems. This framework recognizes abnormal network behaviours predicated on the density characteristics of generated statistical distributions, namely, Dirichlet distributions. The authors illustrate how statistical analysis based on lower-upper interquartile ranges helps in unveiling the normality patterns in the network data being analysed. Consequently, normality patterns allow to flexibly define a suitable statistical model to classify the data at hand, thus making classification decisions more intelligently. A performance evaluation and a comparative study are conducted by using two data sets, showing that the proposed framework outperforms other similar techniques.

Chapter 6 [6] shows the vital role of cybersecurity in e-learning systems, particularly when it comes to ensuring robust and reliable student assessment processes online. Sabbah presents in this chapter two frameworks for overcoming cheating in e-examination processes. The reliability of both systems is tested and validated, showing how both of them outperform existing approaches in terms of risk analysis. A variety of cheating actions and violations are carefully considered in the multi-stage design of the cheating detection system architectures such as impersonation during an e-Examination, messaging patterns and webcam actions. Biometric authentication (e.g. via fingerprints) is adopted as a reliable authentication method to ensure cheating-free examinations.

Classification techniques undeniably play a major role in many cybersecurity-oriented data analytics approaches. Noisy data have long been deemed an influential factor in the accuracy and error rates of classification models. In Chap. 7 [7], Indika revisits a number of popular classification approaches—support vector machines, principal component analytics and random forest ensembles—and presents a comparative study of them, under the perspective of noisy data and its impact on classification accuracy. The author introduces in this chapter a noise removal algorithm and also analyses how skewness, appropriate sample ratios and the chosen classification technique jointly influence the overall classification performance.

Part II—Invited Chapters Besides the regular book chapters outlined above, this book also provides two invited chapters authored by scientists endowed with promising research trajectories in IT security, data analytics and decision support approaches, as well as engagement experience with industry and public sectors to join forces against current cybersecurity challenges.

In Chap. 8 [8], Alamaniotis and Tsoukalas highlight the central role that cybersecurity approaches should play in large-scale smart power systems such as SCADA systems. Power consumption and forecasting information have been identified as key indicators for detecting cyber-attacks in these contexts. Based on this, Alamaniotis and Tsoukalas present an intelligent system method based on Gaussian process regression and fuzzy logic inference, aimed at analysing, reasoning and learning from load demand information in smart grids. The underlying statistical and fuzzy reasoning process allows the system to make highly confident decisions autonomously as to whether current levels of power load demands are legitimated or manipulation by an attacker is taking place.

Garae and Ko provide in Chap. 9 [9] an insightful closure to this book, with a comprehensive overview of analytics approaches based on data provenance, effective security visualization techniques, cybersecurity standards and decision support applications. In their study, Garae and Ko investigate the notion of data provenance as the ability of tracking data from its conception to its deletion and reconstructing its provenance as a means to explore cyber-attack patterns. The authors argue on the potential benefits of integrating data provenance and visualization techniques to analyse data and support decision-making in IT security scenarios. They also present a novel security visualization standard describing its major guidelines and law enforcement implications in detail.

For ease of reference, the table below summarizes the techniques covered and cybersecurity domains targeted by each one of the chapters comprising the book.

	Technique(s) covered	Cybersecurity domain
Chapter 1: Markus Ring et al.	Stream data analytics	Intrusion detection
	Classification, clustering	Insider threats
	Data visualization	
Chapter 2: Philip A. Legg	Human-machine systems	Insider threats
	Decision support	
	Data visualization	
Chapter 3: Irina Măriuca Asăvoae et al.	Mobile computing	Android app collusion detection
	First-order logic	
	Probabilistic models	
	Software model checking	
Chapter 4: Domhnall Carlin et al.	Dynamic opcode analysis	Runtime malware detection
	n-Gram analysis	
Chapter 5: Nour Moustafa et al.	Statistical decision models	Intrusion detection
	Normality analysis	
Chapter 6: Yousef W. Sabbah	Multi-modal data analysis	e-Learning
	Authentication methods	Online examinations
Chapter 7: R. Indika P. Wickramasinghe	Classification (SVM,PCA)	Cybersecurity noisy data removal
	Ensemble classifiers	
	Noisy data management	
Chapter 8: Miltiadis Alamaniotis and Lefteri H. Tsoukalas	Gaussian process regression	Secure smart power systems
	Fuzzy logic inference	
Chapter 9: Jeffery Garae and Ryan K.L. Ko	Data provenance	Security visualization and monitoring
	Data visualization standards	

We would like to thank Springer editorial assistants for the confidence put in this book and their continuous support in materializing it before, during and after its elaboration. We are also very grateful to the authors of selected chapters for their efforts during the preparation of the volume. Without their valuable ideas and contributions, finishing this project would not have been possible. Likewise, we acknowledge all the scientists and cybersecurity experts who generously volunteered in reviewing the chapters included in the book. Finally, we would like to express our special thanks to Dr. Robert McCausland, principal engineer and R&D manager in the Centre for Secure Information Technologies (CSIT), (Queen's University Belfast), for firmly believing in our initiative and strongly supporting it since its inception.

Bristol, UK   Iván Palomares Carrascosa
Belfast, UK   Harsha Kumara Kalutarage
Belfast, UK   Yan Huang
April 2017

References

1. Markus Ring, Sarah Wunderlich, Dominik Grüdl, Dieter Landes, Andreas Hotho. *A Toolset for Intrusion and Insider Threat Detection.*
2. P.A. Legg. *Human-Machine Decision Support Systems for Insider Threat Detection.*
3. I. Măriuca. J. Blasco, T.M. Chen, H.K. Kalutarage, I. Muttik, H.N. Nguyen, M. Roggenbach, S.A. Shaikh. *Detecting malicious collusion between mobile software applications - the Android case.*
4. D. Carlin, P. O'Kane, S. Sezer. *Dynamic Analysis of Malware using Run-Time Opcodes.*
5. N. Moustafa, G. Creech, J. Slay. *Big Data Analytics for Intrusion Detection Systems: Statistical Decision-Making using Finite Dirichlet Mixture Models.*
6. Y. Sabbah. *Security of Online Examinations.*
7. R. Indika P. Wickramasinghe. *Attribute Noise, Classification Technique and Classification Accuracy: A Comparative Study.*
8. M. Alamaniotis, L.H. Tsoukalas. *Learning from Loads: An Intelligent System for Decision Support in Identifying Nodal Load Disturbances of Cyber-Attacks in Smart Power Systems using Gaussian Processes and Fuzzy Inference.*
9. J. Garae, R. Ko. *Visualization and Data Provenance Trends in Decision Support for Cybersecurity.*

Acknowledgements

We would like to extend our gratitude to a number of colleagues, scientists and professionals who willingly helped in reviewing and proofreading this book.

Miltiadis Alamaniotis. Purdue University (United States of America)

Jorge Blasco Alis. Royal Holloway University of London (United Kingdom)

Klavdiya Bochenina. ITMO University (Russia)

Jeremy W. Bryans. Coventry University (United Kingdom)

Roger A. Hallman. SPAWAR Systems Center Pacific (United States of America)

Pushpinder K. Chouhan. Queen's University Belfast (United Kingdom)

Sergey V. Kovalchuk. ITMO University (Russia)

Philip A. Legg. University of the West of England (United Kingdom)

Jing Liao. Hunan University of Science and Technology (China)

Jesús Martínez del Rincón. Queen's University Belfast (United Kingdom)

Robert McCausland. Queen's University Belfast (United Kingdom)

Raghavendra Kagalavadi Ramesh. Oracle Labs Brisbane (Australia)

Jesús Jaime Solano Noriega. University of Occidente (Mexico)

Guillermo Suarez de Tangil. University College London (United Kingdom)

Hui Yu. University of Portsmouth (United Kingdom)

Contents

Part I Regular Chapters

A Toolset for Intrusion and Insider Threat Detection 3
Markus Ring, Sarah Wunderlich, Dominik Grüdl, Dieter Landes,
and Andreas Hotho

**Human-Machine Decision Support Systems
for Insider Threat Detection** 33
Philip A. Legg

**Detecting Malicious Collusion Between Mobile Software
Applications: The Android™ Case** 55
Irina Măriuca Asăvoae, Jorge Blasco, Thomas M. Chen,
Harsha Kumara Kalutarage, Igor Muttik, Hoang Nga Nguyen,
Markus Roggenbach, and Siraj Ahmed Shaikh

Dynamic Analysis of Malware Using Run-Time Opcodes 99
Domhnall Carlin, Philip O'Kane, and Sakir Sezer

**Big Data Analytics for Intrusion Detection System: Statistical
Decision-Making Using Finite Dirichlet Mixture Models** 127
Nour Moustafa, Gideon Creech, and Jill Slay

Security of Online Examinations 157
Yousef W. Sabbah

Attribute Noise, Classification Technique, and Classification Accuracy ... 201
R. Indika P. Wickramasinghe

Part II   Invited Chapters

**Learning from Loads: An Intelligent System
for Decision Support in Identifying Nodal Load Disturbances
of Cyber-Attacks in Smart Power Systems Using Gaussian
Processes and Fuzzy Inference**... 223
Miltiadis Alamaniotis and Lefteri H. Tsoukalas

**Visualization and Data Provenance Trends in Decision Support
for Cybersecurity**.. 243
Jeffery Garae and Ryan K.L. Ko

Contributors

Miltiadis Alamaniotis Applied Intelligent Systems Laboratory, School of Nuclear Engineering, Purdue University, West Lafayette, IN, USA

Jorge Blasco Information Security Group, Royal Holloway University of London, Egham, UK

Domhnall Carlin Centre for Secure Information Technologies, Queen's University, Belfast, Northern Ireland, UK

Thomas M. Chen School of Mathematics, Computer Science & Engineering, City University of London, London, UK

Gideon Creech The Australian Centre for Cyber Security, University of New South Wales Canberra, Canberra, NSW, Australia

Jeffery Garae Cyber Security Lab, Department of Computer Science, University of Waikato, Hamilton, New Zealand

Dominik Grüdl Department of Electrical Engineering and Computer Science, Coburg University of Applied Sciences and Arts, Coburg, Germany

Andreas Hotho Data Mining and Information Retrieval Group, University of Wüzburg, Wüzburg, Germany

Harsha Kumara Kalutarage Centre for Secure Information Technologies, Queen's University of Belfast, Belfast, UK

Ryan K.L. Ko Cyber Security Lab, Department of Computer Science, University of Waikato, Hamilton, New Zealand

Dieter Landes Department of Electrical Engineering and Computer Science, Coburg University of Applied Sciences and Arts, Coburg, Germany

Philip A. Legg Department of Computer Science and Creative Technologies, University of the West of England, Bristol, UK

Irina Mǎriuca Asǎvoae INRIA, Paris, France

Nour Moustafa The Australian Centre for Cyber Security, University of New South Wales Canberra, Canberra, NSW, Australia

Igor Muttik Cyber Curio LLP, Berkhamsted, UK

Hoang Nga Nguyen Centre for Mobility and Transport, Coventry University, Coventry, UK

Philip O'Kane Centre for Secure Information Technologies, Queen's University, Belfast, Northern Ireland, UK

Markus Ring Department of Electrical Engineering and Computer Science, Coburg University of Applied Sciences and Arts, Coburg, Germany

Markus Roggenbach Department of Computer Science, Swansea University, Swansea, UK

Yousef W. Sabbah Faculty of Technology and Applied Sciences, Quality Assurance Department, Al-Quds Open University, Ramallah, Palestine

Sakir Sezer Centre for Secure Information Technologies, Queen's University, Belfast, Northern Ireland, UK

Siraj Ahmed Shaikh Centre for Mobility and Transport, Coventry University, Coventry, UK

Jill Slay The Australian Centre for Cyber Security, University of New South Wales Canberra, Canberra, NSW, Australia

Lefteri H. Tsoukalas Applied Intelligent Systems Laboratory, School of Nuclear Engineering, Purdue University, West Lafayette, IN, USA

R. Indika P. Wickramasinghe Department of Mathematics, Prairie View A&M University, Prairie View, TX, USA

Sarah Wunderlich Department of Electrical Engineering and Computer Science, Coburg University of Applied Sciences and Arts, Coburg, Germany

Part I
Regular Chapters

A Toolset for Intrusion and Insider Threat Detection

Markus Ring, Sarah Wunderlich, Dominik Grüdl, Dieter Landes, and Andreas Hotho

Abstract Company data are a valuable asset and must be protected against unauthorized access and manipulation. In this contribution, we report on our ongoing work that aims to support IT security experts with identifying novel or obfuscated attacks in company networks, irrespective of their origin inside or outside the company network. A new toolset for anomaly based network intrusion detection is proposed. This toolset uses flow-based data which can be easily retrieved by central network components. We study the challenges of analysing flow-based data streams using data mining algorithms and build an appropriate approach step by step. In contrast to previous work, we collect flow-based data for each host over a certain time window, include the knowledge of domain experts and analyse the data from three different views. We argue that incorporating expert knowledge and previous flows allow us to create more meaningful attributes for subsequent analysis methods. This way, we try to detect novel attacks while simultaneously limiting the number of false positives.

1 Introduction

Information security is a critical issue for many companies. The fast development of network-based computer systems in modern society leads to an increasing number of diverse and complex attacks on company data and services. However, company data are a valuable asset which must be authentic to be valuable and inaccessible to unauthorized parties [27]. Therefore, it is necessary to find ways to protect company networks against criminal activities, called intrusions. To reach that goal,

M. Ring (✉) • S. Wunderlich • D. Grüdl • D. Landes
Department of Electrical Engineering and Computer Science, Coburg University of Applied Sciences and Arts, 96450 Coburg, Germany
e-mail: markus.ring@hs-coburg.de; sarah.wunderlich@hs-coburg.de; dominik.gruedl@stud.hs-coburg.de; dieter.landes@hs-coburg.de

A. Hotho
Data Mining and Information Retrieval Group, University of Würzburg, 97074 Würzburg, Germany
e-mail: hotho@informatik.uni-wuerzburg.de

© Springer International Publishing AG 2017
I. Palomares et al. (eds.), *Data Analytics and Decision Support for Cybersecurity*, Data Analytics, DOI 10.1007/978-3-319-59439-2_1

companies use various security systems like firewalls, security information and event management systems (SIEM), host-based intrusion detection systems, or network intrusion detection systems.

This chapter focuses on anomaly-based network intrusion detection systems. Generally, network intrusion detection systems (NIDS) try to identify malicious behaviour on network level and can be categorized into misuse and anomaly detection [25]. Misuse detection utilizes known attacks and tries to match incoming network activities with predefined signatures of attacks and malware [14]. Consequently, only known attacks can be found and the list of signatures must be constantly updated [18]. The increasing trend of insider attacks complicates this challenge even more. It is harder to identify signatures that indicate unusual behaviour as these behaviours may be perfectly normal under slightly different circumstances. Anomaly detection systems on the other hand assume that normal and malicious network activities differ [18]. Regular network activities are modelled by using representative training data, whereas incoming network activities are labelled as malicious if they deviate significantly [14]. Thus, anomaly detection systems are able to detect novel or obfuscated attacks. However, operational environments mainly apply misuse detection systems [52]. Sommer and Paxson [52] identify the following reasons for the failure of anomaly-based intrusion detection systems in real world settings:

1. high cost of false positives
2. lack of publicly available training and evaluation data sets
3. the semantic gap between results and their operational interpretation
4. variability of input data
5. fundamental evaluation difficulties

In this contribution, we propose a novel approach for anomaly based network intrusion detection in which we try to consider the challenges identified by Sommer and Paxson [52]. We report on our ongoing work that aims to develop an interactive toolset which supports IT security experts by identifying malicious network activities. The resulting toolset Coburg Utility Framework (CUF) is based on a flexible architecture and offers a wide range of data mining algorithms. Our work addresses various aspects that may contribute to master the challenge of identifying significant incidents in network data streams, irrespective of their origin from inside or outside of a company network. In particular, our approach builds upon flow-based data. Flows are meta information about network communications between hosts and can be easily retrieved by central network components like routers, switches or firewalls. This results in fewer privacy concerns compared to packet-based approaches and the amount of flow data is considerably smaller in contrast to the complete packet information.

The resulting approach is multistaged: We first propose an enrichment of flow-based data. To this end, we collect all flows within a given time window for each host and calculate additional attributes. Simultaneously, we use additional domain knowledge to add further information to the flow-based data like the origin of the *Source IP Address*. In order to detect malicious network traffic, the enriched flow-based data is analysed from three different perspectives using data mining algorithms. These views are adjusted to detect various phases of attacks like *Scanning* or *Gaining Access* (see Sect. 3.1.1). We are confident that collecting flows for each host separately for a certain time window and the inclusion of domain knowledge allows us to calculate more meaningful attributes for subsequent analysis methods. Further, the three different analysis views allow us to reduce the complexity in each single view. We also describe our process to generate labelled flow-based data sets using *OpenStack* in order to evaluate the proposed approach.

The chapter is organized as follows: The next section introduces the general structure of our toolset Coburg Utility Framework (CUF). Section 3 proposes our data mining approach to analyse flow-based data streams using CUF. Then, the generation of labelled flow-based data sets for training and evaluation is described in Sect. 4. Section 5 discusses related work on flow-based anomaly detection. The last section summarizes the chapter and provides an outlook.

2 Coburg Utility Framework

We use the following definitions: A data stream $X_S = \{X_1, X_2, \ldots, X_n\}$ is an unbounded set ($n = \infty$) of data points X_i. Each data point X_i is characterized by a set of attributes. A data set X_D consists of a fixed number of data points X_i. When talking about network data, each flow can be considered a data point.

The objective of our research is a methodology for analysing flow-based data streams to detect malicious network activities. To this end, a combination of various data mining methods (clustering, classification and visualization) needs to be explored. Different configurations are then compared and evaluated with respect to their quality. An appropriate workflow can only be found by experimenting with different processing steps on real world data. By assessing the results we are able to figure out the best possible configuration. Hence, a highly flexible experimental environment is required to allow for an easy setup of different methods. Hereby, a wide range of tools for modelling, visualization and evaluation is used.

The Coburg Utility Framework (CUF) aims to fulfil these requirements. The underlying core architecture of CUF is presented in the following section. Section 2.2 provides an overview of available algorithms for offline and online analysis. CUF also incorporates offline algorithms which we developed and applied in earlier work [27] since these algorithms can be used for further and deeper investigations of malicious network activities.

2.1 Architecture of CUF

CUF implements a pipes-and-filters architecture which is often used in software applications that handle and process data streams. Filters constitute independent

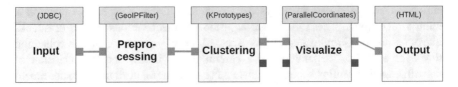

Fig. 1 Representation of a data mining workflow in CUF

working steps that manipulate incoming data. Pipes connect pairs of filters and pass data on to other filters. The architecture of CUF ensures optimal encapsulation of pipes and filters, and both are realized as services with a common interface. Services are implemented by a service provider and loaded dynamically by a service loader. In addition, this architecture provides an easy way of integrating new filters, e.g., when integrating a new clustering algorithm as a filter, all other filters and pipes stay unchanged.

As an enhancement to the pure pipes-and-filters pattern, CUF allows to split and merge workflows. Thus, different clustering filters or multiple instances of one filter with different parameter settings can be combined to process data in a single run and compare their results directly. Since each of these steps is implemented as an individual filter in CUF, different workflows can easily be set up and executed.

Figure 1 shows a simple data mining workflow in CUF to cluster network data. At first, an input filter reads the data from a database. Then, a preprocessing filter adds additional information to each data point. In this case, the *GeoIPFilter* adds to each data point the corresponding geographical coordinates of the *Source IP Address* and *Destination IP Address* using an external database. The third filter in the processing chain is the clustering algorithm *k-Prototypes*. This filter sorts the incoming data points in a predefined number of k clusters (groups) according to their similarity. The fourth filter visualizes the results in form of *Parallel Coordinates* and the last filter writes the results to disk. Input and output interfaces of the filters are represented by the coloured rectangles in Fig. 1. Input interfaces are on the left side and output interfaces are on the right side. Green rectangles transport data points, whereas blue rectangles transport cluster objects. Figure 1 shows that the *Parallel Coordinates* filter is able to process data points (green input rectangle) or cluster objects (blue input rectangle). The ability of filters to read and generate different output formats allows us to easily split and merge workflows.

One of the major challenges in analysing network data is the large amount of data generated by network devices. Company networks generate millions of network flows per hour. Consequently, an efficient and fast processing chain is required. Here, the pipes-and-filters architecture of CUF itself has a big advantage. Since filters work independently, each of them is executed in its own thread such that multicore architectures may easily be utilized to reduce execution time.

2.2 Filters in CUF

This section provides an overview of available filters in CUF. We implement many filters on our own to increase flexibility and customization of algorithms for flow-based data. For example, we implemented the *k-Prototypes* algorithm to be able to use adjusted distances measures. When no such adjustments are necessary, we prefer the use of publicly available implementations. For instance, we include several classifiers from the WEKA[1] toolkit like *J48*.

We distinguish five subcategories of filters, namely *Input and Output, Preprocessing, Clustering, Classification* and *Evaluation*.

2.2.1 Input and Output

As the name suggests, input and output filters are responsible for reading and writing data. CUF offers various input filters which can read data from different sources like text files, binary files, or databases. For each input filter, a corresponding output filter is available for storing data in a particular format. Binary files are primarily used to read and write temporary results from clusterings. For text files, CUF offers two formats: *CSV* and *HTML*. The *CSV* format is most widely used. Yet, analysing *CSV* files with many columns and lines can quickly become confusing. Therefore, CUF may also write data in a formatted table and store it in *HTML* format.

We are primarily interested in analysing flow-based data streams. In experimental settings, however, network data streams are often not available. Instead, flow-based data streams are recorded from network devices and stored in *CSV* files. Each recorded flow (data point) contains an attribute named *Date first seen*. This attribute depicts the timestamp at which the corresponding network device created the flow. For simulating live data streams from these stored *CSV* files, CUF offers an additional stream simulation filter. The stream simulation filter emulates a data stream by extracting the attribute *Date first seen* from each flow and calculates the time difference between two successive flows. The filter forwards the flows with respect to the calculated time differences in the processing chain to simulate real traffic.

2.2.2 Preprocessing

CUF contains a wide range of preprocessing filters which, in most cases, are not limited to network data. Preprocessing filters clean data (e.g. by removing inconsistent data), transform attribute values, or add further information. Some preprocessing filters use explicit domain knowledge to add further information to data points.

---

[1]http://www.cs.waikato.ac.nz/ml/index.html.

An important aspect in network data analysis is heterogeneous data, namely data points that are composed of continuous and categorical attributes. Continuous attributes take real numbers as values and thus they have a natural order in their value range. This allows simple calculations of similarities between different values. Examples for continuous attributes are the *bytes* or the *timestamp* of a network flow. The values of categorical attributes are discrete and have no natural order in their value range. This makes it hard to calculate similarities between categorical values, since simple distance measures like the *Euclidian distance* can not be applied. Examples for categorical attributes are the *Source IP Address* or *Source Port* of a flow. The mixture of continuous and categorical attributes complicates matters since most data mining algorithms can only handle either continuous or categorical attributes. To transform the problem of heterogeneous data into a well-known problem, CUF offers various preprocessing filters to discretize continuous attributes (e.g. the *Hellinger Discretization* [28] or the *Multi-Interval Discretziation using Minimal Description Length* [15]) or to transform categorical attributes to continuous attributes (e.g. a transformation to binary attributes for each categorical value).

Further, CUF contains a novel distance measure *ConDist* [44]. *ConDist* [44] is able to calculate distances between data points which contain both, continuous and categorical attributes. *ConDist* utilizes the *Minkowski distance* to calculate the distance for continuous attributes. For categorical attributes, *ConDist* uses correlated context attributes to calculate distances. It also considers the quality of information that can be extracted from the data set and automatically weights the attributes in the overall distance calculation.

2.2.3 Clustering

In general, the first goal of any type of data analysis is a better understanding of the data [27]. Clustering is an unsupervised technique, meaning it uses no a priori knowledge about the data. Consequently, appropriate clustering techniques can only be determined experimentally. In order to avoid restricting the range of techniques, CUF integrates clustering algorithms from all categories, namely partitioning (k-means and variants), hierarchical (Lance-Williams [33], ROCK [20] and extensions), grid-based and density-based algorithms (CLIQUE [2] and extensions) [27]. Any clustering algorithm may choose an appropriate distance measure, ranging from *Minkowski distances* for continuous data over the *Jaccard Index* for categorical attributes to *ConDist* [44] for heterogeneous data. The latter is even capable of self-adapting to the underlying data set.

Further, CUF integrates the stream clustering algorithms *CluStream* [1] and *DenStream* [6] both of which are based on the principle of micro and macro clusters. An online component clusters incoming data points into micro clusters. The offline component is triggered by the user, uses micro clusters as input data points, and presents a clustering result for a certain timeframe to the user. *CluStream* [1] and *DenStream* [6] can only process continuous attributes in their standard form.

Therefore, CUF also integrates *HCluStream* [64] and *HDenStream* [29] which have been proposed as extensions in order to handle categorical attributes in addition to numerical ones.

2.2.4 Classification

Classification is a basic task in data mining and used to automatically assign (classify) data points into predefined groups (classes). Such a data mining model takes as input a set of labelled examples. CUF integrates different classifiers like *Neural Networks, Decision Trees, k-Nearest-Neighbour* or *Support Vector Machine*. Classification proceeds in two steps. In the first step, the classifier is trained using a labelled set of training data. The training data set contains a fixed number of class information. Each training data point X_i is composed of a constant set of attributes and a label indicating the class to which the data point belongs. The classifier learns the characteristics of the different classes and builds a model or function for the assignment (classification) of unlabelled data points. In the second step, the classifier uses this model or function to classify new (unseen) data points [56].

2.2.5 Evaluation

CUF provides a wide range of filters for result evaluation. First of all, filters are integrated to calculate default evaluation measures like *Accuracy, F1-Score, Recall* or *Precision*. However, these measures can only be calculated when a ground truth is available, like the labels in the classification setting. If there is no ground truth available, CUF offers filters which calculate intrinsic validation measures. Regarding the evaluation of data stream clustering, Hassani and Seidl examine the performance and properties of eleven internal clustering measures in [21]. Since Calinski-Harabasz [5] emerges as best internal evaluation measure it is also implemented in CUF. Besides Calinski-Harabasz, other promising evaluation methods are implemented as well (e.g. CPCQ-Index [30], CDbw-Index [9], or Davies-Bouldin-Index [13]). These intrinsic cluster validation measures evaluate the clustering results by their compactness and separability. However, these key measurements can only provide an overall assessment of the results and give no further insights.

Therefore, CUF provides various visualization filters for deeper investigations. *Data plots* (see Fig. 2) may give a first impression of the data. *Bar diagrams* give an overview of the number of generated clusters and their number of data points. Further, CUF may use parallel coordinates, pixel-based visualization, or radial visualizations to display large amounts of multi-dimensional data to a human security expert. Figure 3 shows the visualization of a network data set in parallel coordinates as well as in radial coordinates. In parallel coordinates, each attribute is displayed on an axis. All axes are displayed parallel to each other on the screen. Each data point is represented as a line from left to right. The situation is slightly

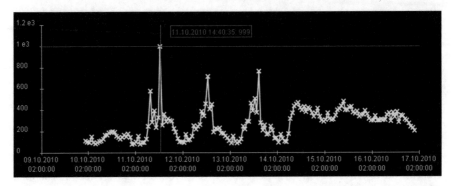

Fig. 2 Data plot which represents the number of flows per time

A. Parallel Coordinates

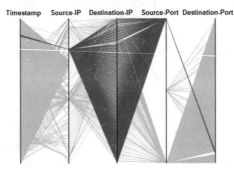

B. Radial Coordinates

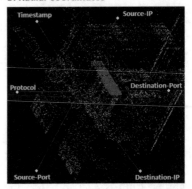

Fig. 3 Representation of a network data set with a *horizontal Port Scan* in parallel coordinates (**a**) and a *vertical Port Scan* in radial coordinates (**b**). The flows which belong to the *Port Scan* are highlighted in *red*

different for radial coordinates where each attribute is displayed on an axis as well, but data are arranged in a circular layout. However, it has to be taken into account that opposing axis are influencing each other. Other visualizations may be used to get an overview of or detailed insights into the overall result.

This way, we tackle the fundamental challenge of evaluation of anomaly based intrusion detection systems by using a broad range of evaluation technologies.

3 Online Analysis of Network Data Streams

In this section, we introduce our approach of analysing flow-based data streams with CUF. First, we investigate the general problem setting and deduce some implications for our approach in Sect. 3.1. Section 3.2 gives an overview of the proposed approach and provides the underlying ideas. Then, the integration of additional domain knowledge is described in Sect. 3.3. Sections 3.5–3.7 describe the three different views in which the incoming flow-based data stream is analysed.

3.1   General Problem Setting

The objective of our research is to identify malicious network traffic in flow-based data streams using data mining methods. Hence, we now discuss typical attack scenarios, available data and necessary data preparation steps.

3.1.1   Attack Scenarios

Attack scenarios often follow predefined patterns. Literature usually distinguishes various phases of attacks. A popular definition by Skoudis and Liston [51] is depicted in Fig. 4 and contains the following phases:

1. Reconnaissance
2. Scanning
3. Gaining Access
4. Maintaining Access
5. Covering Tracks

In each phase, different methods are used to reach the corresponding goals.

In the first phase (*Reconnaissance*), the attacker tries to gather as much information as possible about the target system. Skoudis and Liston [51] identify various *Reconnaissance* techniques, e.g. *web searches, whois database analysis, domain name systems, social engineering* or *dumpster diving*. A popular starting point in this phase would be to search the companies website. In case of *social engineering*, names of the current CEOs or other employees (e.g. server administrators) would be targeted. Attackers commonly use such information by acting as an administrator via email to get critical data from a trustful employee. *Domain name systems*

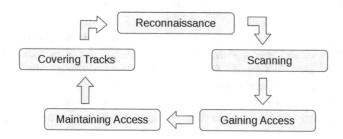

Fig. 4 Overview of the five attack phases

can be used to identify concrete IP ranges. Alternatively, it is possible to retrieve usernames, (inital) passwords and internal *IP Addresses* via *dumpster diving*.

Scanning is the second phase and uses information which was gathered during *Reconnaissance* to examine the target system. In essence, the *Scanning* phase is an extension of *Reconnaissance*. Directly accessible hosts can be identified by *IP Range Scans*. Then, *Port Scans* are used to identify open ports and thereby potential attacking points. A variety of powerful scanning tools such as *nmap*[2] exists. Further, these scanning tools contain features for deeper analysis like operation system detection.

In the third phase (*Gaining Access*), the actual attack takes place. Typical attacks in this phase are *Denial of Service* (*DoS*), *SQL injections*, *PHP exploits*, or *Brute Force attacks*. Exemplary, *DoS* attacks try to make the target system unavailable by flooding the target with illegitimate packets [8], while *SSH Brute Force* attacks are used to gain access to a remote host by repeatedly trying to guess the correct login information.

After gaining access, attackers usually want to maintain it. This can be accomplished in phase four (*Maintaining Access*) via typical techniques like *trojan horses*, *bots*, *backdoors* or *rootkits* [51]. The compromised system can be used as starting base for deeper exploration of the target network and to collect information for further attacks.

The *Covering Tracks* phase completes the cycle of the attack scenario. Attackers typically prefer to stay hidden, so they can maintain control of the systems as mentioned in phase four. Staying hidden buys time for the attackers and enables them to steal data, consume CPU cycles, launch other attacks or just maintain the control for future actions [51].

Attack scenarios do not necessarily include all phases. For example, an insider (e.g. an employee) already holds information that an outsider would acquire in phase one. Consequently, insiders would start at phase two or even three. In addition, not all attackers are interested in keeping access and wiping out their traces. In these cases the phases four and five are optional.

3.1.2 Flow-Based Data

As already mentioned above, we focus on flow-based data for intrusion and insider threat detection. To reiterate, flows contain meta information about connections between two network components. A flow is identified by the default five tuple: *Source IP Address, Source Port, Destination IP Address, Destination Port* and *Transport Protocol*. We capture flows in unidirectional *NetFlow* format [10] which typically contains the attributes shown in Table 1. These attributes are typical in flow-based data and also available in other flow standards like *IPFIX* [11] or *sFlow* [39]. *NetFlow* terminates a flow record in two cases: (1) a flow receives no

---

[2]https://nmap.org/.

Table 1 Overview of the used *NetFlow* attributes in our approach. The third column gives a short description of the attributes

Nr.	Name	Description
1	Src IP	Source IP address
2	Src port	Source port
3	Dest IP	Destination IP address
4	Dest port	Destination port
5	Proto	Transport protocol (e.g. ICMP, TCP, or UDP)
6	Date first seen	Start time flow first seen
7	Duration	Duration of the flow
8	Bytes	Number of transmitted bytes
9	Packets	Number of transmitted packets
10	Flags	*OR* concatenation of all TCP Flags

data within α seconds after the last packet arrived (inactive timeout) or (2) a flow has been open for β seconds (active timeout). By default, *NetFlow* uses the values $\alpha = 15$ and $\beta = 1800$.

NIDS usually operate either on flow-based data or on packet-based data. We analyse flow-based data for several reasons. In contrast to packet-based data, the amount of data can be reduced, fewer privacy concerns are raised, and the problem of encrypted payloads is bypassed. Further, problems associated with the variability of input data [52] are avoided. Flows have a standard definition and can be easily retrieved by central network components. Another advantage of flows is that huge parts of the company network can be observed. For example, firewall log files would limit the analysed data to the traffic which passes the firewall. Consequently, insider attacks do not appear in these log files as they usually do not pass the firewall. In contrast to that, traffic between internal network components always passes switches or backbones. In conclusion, the use of flows allows to analyse the whole company network traffic independently of its origin.

Flows come as either unidirectional or bidirectional flows. Unidirectional flows aggregate those packets from host A to host B into one flow that have identical five tuples. The packets from host B to host A are merged to another unidirectional flow. In contrast, bidirectional flows contain the traffic from host A to host B as well as vice versa. Consequently, bidirectional flows contain more information. However, we decided to use unidirectional flows since company backbones often contain asymmetric routing [23] which would distort the information in bidirectional flows.

The individual phases of an attack (see Sect. 3.1.1) have different effects on flow-based data. The *Reconnaissance* phase has no influence on the data since methods like dumpster diving or social engineering generate no observable network traffic within the company. In comparison, the other four phases generate observable network traffic.

It should be noted that the detection of attacks using host-based log files could sometimes be easier than the analysis of flows, e.g. failed *ssh logins* are stored in the ssh-log-file. However, regarding phase five (*Covering Tracks*), attackers usually manipulate the log files on the host to wipe their traces. Since we use flow-based data of network components and no host-based log files, the covering of tracks fails in our anomaly based intrusion detection system. This would only be possible if the attacker hacks the network components and manipulates the flows.

3.1.3 Data Preparation

We follow the typical phases of the CRISP (Cross-Industry Standard Process for Data Mining) model [49] which defines a standard model for Data Mining workflows. Business understanding and data preparation are necessary prerequisites for the application of Data Mining methods. In our setting, business understanding corresponds to the analysis of hacking phases (Sect. 3.1.1) while data understanding complies with the analysis of flow-based data (Sect. 3.1.2).

The next step in the workflow is the data preparation before data mining methods can be applied. Reasons are the need for attributes which represents hidden information or the requirement of some data mining methods, as they are only able to handle some types of attributes. For example, when using *Neuronal Networks* the input data should consist of exclusively continuous attributes. Hence, categorical attributes like *IP Addresses* have to be transformed. Several aspects have to be considered when preprocessing flow-based data. Clustering (Sect. 2.2.3), classification (Sect. 2.2.4) and outlier detection algorithms are usually based on a definition of distance or similarity. Applying these algorithms directly on the attributes of flow-based data (see Table 1) could lead to fatal problems. Therefore, it is necessary to abstract certain attributes of the flow in a preprocessing step. For example, when facing the same attack twice, but from a different *Source IP Address* and on a different *Destination IP Address* to a later time, at least three of the ten attributes of a flow will differ. As a consequence, the flows of the two attacks are dissimilar and categorized to different groups.

3.1.4 Implications

Considering our above analysis of the general problem setting, where we investigated attack scenarios, the underlying flow-based data and necessary data preparation steps, we can draw several conclusions:

1. A basic assumption is that due to the big number of various applications and services it is nearly impossible to decide if a single flow is normal or malicious traffic based on the available attributes (see Table 1). This assumption is supported by the fact that normal user behaviour and malicious user behaviour are characterized by sequences of flows. Lets illustrate this by an example. Assume a *Vertical Port Scan* attack. Here, the attacker scans some or all open ports on a target systems [55]. Since *Source Port* and *Destination Port* are keys

of the default five tuple for creating flows, each scanned port generates a new flow. Another example is the loading of a web page. Often different pictures are reloaded or other web pages are included. In such cases, the client opens various *Source Ports* or sends request to different web servers (different *Destination IP Addresses*). For these reasons, it makes more sense to collect multiple flows for each host rather than to analyse each flow separately.

2. Different attack phases have different effects on flow-based data. In the second phase (*Scanning*) different services and/or hosts are targeted, whereas in the third phase (*Gaining Access*), a concrete service of a host is attacked. In the fourth phase (*Maintaining Access*), flow characteristics like transmitted *Bytes* or *Packets* seem to be normal, only the origin of the connections are suspicious. Consequently, it makes more sense to analyse the flow-based data from different views to detect different attack phases.

3. The information within flow-based data is limited. Therefore, flows should be enriched with as much information about the network as possible using domain knowledge.

3.2   Outline of the Proposed Approach

This section provides an overview of the proposed approach and discusses the underlying ideas. Figure 5 shows the essential components of our approach.

The *IP Address Info Filter* is the first filter in the processing chain. It receives the flow-based data stream X_S from central network components and incorporates domain knowledge about the network (see Sect. 3.3). Data is passed through the *Service Detection Filter* (Sect. 3.4) to identify the services of each flow (e.g. *SSH*, *DNS* or *HTTP*). The *Collecting Filter* is the central component of our approach. It also incorporates domain knowledge about the network and receives the enriched flow-based data stream X_S from the *Service Detection Filter*. Based on the observations above, in its first step, the *Collecting Filter* collects all incoming flows for each user separately. User identification is based on the *Source IP Address* of the flow. A parameter δ controls the windows-size (in seconds) of flows which are collected for each user. The larger the parameter δ, the more memory is necessary but in consequence, the quality of the calculated summary of flows increases. The *Collecting Filter* creates one *Network data point* for each user and each time window. For each user and identified service within the time window a *Service data point* and *User data point* is created. Each of these data points is created for investigating the user flows from a specific view, namely *Network Behaviour Analyser*, *Service Behaviour Analyser* and *User Behaviour Analyser* in Fig. 5. The *Network data point* contains specific information about the users' network behaviour and is described in Sect. 3.5. The *Service data point* contains specific information about the usage of the concrete service and is described in Sect. 3.6. The *User data point* contains specific information, e.g. if the behaviour

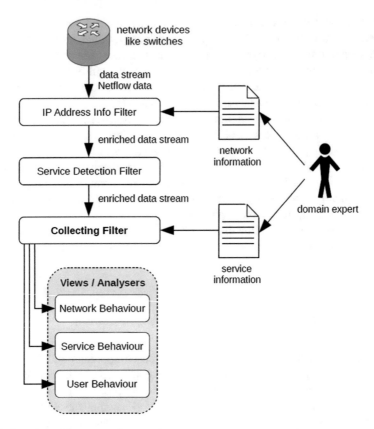

Fig. 5 Overview of the proposed approach

is typical for a user or not. It is described in Sect. 3.7. We argue that incorporating domain knowledge from IT security experts and other flows allow us to create more meaningful attributes for downstream analysis methods.

The *IP Address Info Filter*, *Service Detection Filter*, *Collecting Filter*, *Network Behaviour*, *Service Behaviour* and *User Behaviour* are implemented as separate filters in CUF which can be run independently and in parallel.

3.3   Integration of Domain Knowledge

Arguably, the performance of a NIDS increases with the amount of domain specific information about the network. Therefore, we integrate more detailed information about *Source IP Address*, *Destination IP Address*, *Source Port* and *Destination Port* in our system.

```
#Subnet IP Address, Subnetwork, Organization, isIntern, isServer
192.168.1.0 , 16, General    , 1, 0
192.168.1.0 , 24, Management, 1, 0
192.168.2.0 , 24, Developer , 1, 0
192.168.3.0 , 24, Server     , 1, 0
192.168.3.2 , 32, MailServer, 1, 1
192.168.3.3 , 32, FileServer, 1, 1
192.168.3.4 , 32, WebServer , 1, 1
192.168.2.42, 32, PrinterMan, 1, 1
192.168.3.42, 32, PrinterDev, 1, 1
```

Fig. 6 Exemplary configuration file for integrating domain knowledge

First, we integrate additional information about the *Source IP Address* and *Destination IP Address* through the *IP Address Info Filter*. Therefore, the network administrator has to set up a *csv* file with all network addresses of the company. Further, the administrator has to add each *IP Address* of internal servers. Figure 6 shows such a sample *csv* file.

The first column in Fig. 6 is a specific *IP Address* or the address of an internal subnet. The second column defines the size of the subnet in column one, where 32 means that this entry refers to a single *IP Address*. The third column describes the organization of the subnet or the function of the particular *IP Address*. The last two columns indicate if the *IP Address/Range* is internal and if the *IP Address/Range* is a server. Normally, all servers are part of the company network and marked as internal. However, the configuration file allows the administrator to add special roles to extern but known servers (e.g. servers of other companies for joint projects). Any *IP Address/Range* that can not be matched to a given entry within the configuration file is treated as an external address. Based on this information, the *IP Address Info Filter* adds the corresponding organization and the two flags *isIntern* and *isServer* to each *Source IP Address* and *Destination IP Address*.

The second configuration file contains information about used ports in the company. Here, the network administrator has to flag open ports of internal servers, e.g. *port* 20 and 21 for *FTP* or *port* 22 for *SSH*. By default, all *ports* between 0 and 1023 (the standardized ports) are considered as open ports. Requested connections to non open ports increase the level of suspiciousness. This configuration file is read by the *Collecting filter* for identifying clients and servers within a flow.

3.4 Service Detection Filter

The *Service Detection Filter* classifies each flow with respect to their services (e.g. *HTTP*, *SSH*, *DNS* or *FTP*). Right now, this filter uses a common identification method which is based on evaluating known port numbers assigned by the *Internet Assigned Numbers Authority* (*IANA*).[3]

Unfortunately, this approach is no longer viable because many applications do not use fixed port numbers [65]. Another problem when evaluating known port numbers is that many applications tunnel their traffic through port 80 (e.g. Skype).

Therefore, we intend to integrate more sophisticated service detection algorithms in the future. Nguyen et al. [37] and Valenti et al. [59] broadly review traffic classification using data mining approaches which is not limited to flow-based data. Several approaches for flow-based service classification have been published [32] and [65]. Moore and Zuev [32] show the effectiveness of a Naive Bayes estimator for flow-based traffic classification. Zander et al. [65] use *autoclass*, an unsupervised Bayesian classifier which learns classes inherent in a training data set with unclassified objects. A more recent approach of service classification using *NetFlow* data is given by Rossi and Valenti [46].

3.5   Network Behaviour Analyser

The *Network Behaviour Analyser* evaluates hosts with respect to their general network behaviour. Consequently, this analyser primarily checks if the number and the kind of connections are normal or suspicious for a specific host. The primary goal is to identify activities in the *Scanning* phase (see Sect. 3.1.1). Therefore, the main attacks in this scenario are *IP Range Scans* or *Port Scans*. For the detection of *Port Scans*, a more detailed analysis is required. *Port Scans* can be grouped into *horizontal scans* and *vertical scans*. In case of more common *horizontal scans* the attacker exploits a specific service and scans numerous hosts for the corresponding port [55]. In contrast, *vertical scans* target some or all ports of a single host.

Since the scanning behaviour of attacker and victim hosts are different for *TCP* and *UDP*, they need to be treated separately. The most common *TCP* scan is the *SYN-scan*. In this case, the attacker sends the initialization request of the 3-Way-Handshake. If the port is open, a *SYN-ACK*-response is sent by the victim. Otherwise, the victim host responds with a *RST-Flag*. It should be noted that there are different approaches of *TCP* scans, e.g. sending a *FIN* flag instead of the initialization *SYN* flag for bypassing firewall rules. These approaches are described in more detail in the documentation of the popular *nmap*[4] tool.

Scanning *UDP* ports differs fundamentally from scanning *TCP* ports. Successful addressing *UDP* ports does not necessarily render a response. However the same behaviour can be observed if a firewall or other security mechanisms blocks

---

[3]http://www.iana.org/assignments/service-names-port-numbers/service-names-port-numbers. xhtml.

[4]https://nmap.org/.

the request. If the attacker addresses a closed *UDP* port, the victim sends an *ICMP* unreachable message in return. However, most operating systems limit the number of *ICMP* unreachable messages to one per second.

In consequence, it is easier to detect the targeted victim instead of the attacker, since victims follow rules predefined by protocols. In contrast, attackers may vary their behaviour regarding the protocol to trick the systems. Also, it is more likely to detect the scanning of closed ports due to the atypical behaviour of request-response. Due to these observations, data points are calculated in the *Collecting Filter* with corresponding attributes for this view. The calculated attributes contain values like the number of flows within a time window, the number of sent or received *RST-*Flags, the number of sent or received *ICMP* unreachable messages as well as the number of requested ports per *Destination IP Address*. Since we consider both, the view of the victim and the attacker, it is easier to detect distributed *Port Scans* too.

In preliminary experiments, we applied various classifiers (*J48 Decision Tree, k-Nearest-Neighbour, Naive Bayes* or *SVM*) for the detection of *IP Range Scans* and *Port Scans*. The proposed approach seems to work on our emulated data sets. We will describe the process of data emulation in Sect. 4.

Several other methods which use similar approaches for *Port Scan* detection are discussed in literature. Methods like Time-based Access Pattern, and Sequential hypothesis testing (TAPS) [54] use the ratio of *Destination IP Addresses* and *Destination Ports* to identify scanners. If the ratio exceeds a threshold, the host is marked as scanner. Threshold Random Walk (TRW) [24] assumes that a scanner has more failed connections than a legitimate client. For identification of failed connections, TRW also evaluates the *TCP-Flags*.

3.6   Service Behaviour Analyser

The *Service Behaviour Analyser* evaluates the hosts from the view of their correct usage of services. Consequently, the main goal of this analyser is to check if the use of the current service is normal or malicious for a host. Primary target is to recognize the *Gaining Access* phase mentioned in Sect. 3.1.1. *DoS* or *SSH Brute Force* attacks are typical representatives.

For the detection of misused services, it is necessary to collect all flows of this service within the time window. Therefore, we use the service attribute added by the *Service Detection Filter* (Sect. 3.4). All flows of the host within a time window which share the same *service* and the same *Destination IP Address* are collected. Based on these collected flows, the *Collecting Filter* calculates data points with adjusted attributes for this view. The calculated attributes contain values like the sum of transmitted *Bytes* and *Packets*, the duration of the flows, or the number of flows. More useful attributes like the number of open connections or the number of successfully closed connections can be derived from *TCP Flags* (if available). The *Collecting Filter* also builts attributes which give additional information about the source and destination using the domain knowledge of Sect. 3.3.

Again, we applied various classifiers (*J48 Decision Tree*, *k-Nearest-Neighbour*, *Naive Bayes* or *SVM*) for the detection of misused services. Preliminary experiments regarding the detection of *SSH Brute Force* attacks on our emulated data sets (Sect. 4) show promising results.

Quite a few similar approaches have already been published. For instance, Wagner et al. [60] proposed a kernel for anomaly detection, especially for *DDoS* attacks. Their kernel includes all recorded flows within a time window and considers the *IP Addresses* and *Bytes* of the flows. Najafabadi et al. [34] propose an approach for detecting *SSH Brute Force* attacks using aggregated *NetFlow* data. The authors incorporate domain knowledge about *SSH Brute Force* attacks, extract discriminating attributes from *NetFlows* and use classifiers for the detection. Hellemons et al. [22] propose *SSHCure*, a flow based SSH intrusion detection system. The authors analyse the model of *SSH Brute Force* attacks and derive attributes for their detection in *NetFlow* data. However, unlike these approaches, our approach is more general and not limited to one specific source.

3.7   User Behaviour Analyser

The *User Behaviour Analyser* filter evaluates the hosts with respect to their used services being typical. The main goal of this analyser is to recognize if the current connection is normal or malicious for this user and to identify already infected and misused hosts. Primary target is to recognize the *Maintaining Access* and *Covering Tracks* phases mentioned in Sect. 3.1.1.

Once a server is infected, the attacker knows a valid combination of username and password and is able to start a *SSH* session to that server. In this case, the traffic characteristics like transmitted *bytes*, *packets* or *duration* seem to be legitimate. Therefore, the *Collecting Filter* calculates data points for this view which strongly consider the source, destination and services of the flows. Here, the domain knowledge of Sects. 3.3 and   3.4 is used. For example, the calculated attributes describe if the *Source (Destination) IP Address* is internal or external, if the *Source (Destination) IP Address* is a server or a client, and which organizations (see Fig. 6) the *Source (Destination) IP Address* belongs to. Further, the identified service by the *Service Detection Filter* is added as an additional attribute.

Right now, we use a simple rule learner which generates rules regarding normal and malicious behaviour. If an unknown combination occurs, the corresponding connection information is sent to the domain experts for further investigation. Rules generated in this case would look like the following:

$$organization_{source} = extern \land organization_{destination} = server \land service = SSH \rightarrow Malicious$$

This example would imply malicious behaviour, since we would not expect a valid SSH connection to an internal server from outside the company network.

4   Data Generation

Labelled publicly available data sets are necessary for proper comparison and evaluation of network based intrusion detection systems. However, evaluation of our system proves to be difficult due to the lack of up-to-date flow-based data sets. Many existing data sets are not publicly available due to privacy concerns. Those which are publicly available often do not reflect current trends or lack certain statistical characteristics [50]. Furthermore, correct labelling of real data proves to be difficult due to the massive and in-transparent generation of traffic in networks. In order to overcome these problems, we create labelled flow-based data sets through emulation of user activities in a virtual environment using OpenStack [45].

In this section, some prominent existing data sets are presented as well as our own approach to generate data sets for IDS evaluation.

4.1   Existing Data Sets

DARPA98 and DARPA99 from the MIT Lincoln Laboratory were among the first standard packet-based data sets published for evaluation purposes. Those data sets were created by capturing simulated traffic of a small US Air Force base with limited personnel via tcpdump [26]. The MIT Lincoln Laboratory also provided the KDD CUP 99 data set, which is a modified version of the DARPA98 data set [17]. Each data point of the KDD data set consists of 41 attributes. The KDD data set, however, has a few problems, one being the huge number of redundant records. To overcome these problems, the NSL-KDD data set was generated by Tavallaee et al. [57]. Since publicly available data sets are sparse due to privacy concerns, a lot of today's work is based on the DARPA [40, 47] and KDD [7, 12, 41] data sets. As DARPA data sets were created more than 17 years ago, it is questionable if they still reflect relevant up-to-date scenarios appropriate for IDS evaluation [19, 62, 66].

Besides the data sets being outdated, conversion of packet-based to flow-based data sets turns out to be tricky, if the data set is not available in a standard packet-based format like *pcap*. Since we prefer to analyse network flows naturally flow-based data sets would be best. Sperotto et al. [53] created one of the first publicly available flow-based labelled datasets by monitoring a single honeypot. Due to the traffic being recorded by monitoring a honeypot, the data set mainly consists of malicious data. Thus, the detection of false positives could not be determined during the evaluation [53]. For a more comprehensive IDS evaluation, a more balanced data set would be preferable.

In 2014, Wheelus et al. [62] presented the SANTA dataset which consists of real traffic as well as penetration testing attack data. The attack data was labelled via manual analysis [62]. In 2015, Zuech et al. [66] introduced a two part dataset named IRSC. The IRSC set comprises a netflow data set as well as full packet capture set. The data sets were labelled manually for uncontrolled attacks and via an IP filter for controlled attacks [66]. Although the two flow-based data sets [62, 66] are up-to-date, they are not publicly available as of now.

Another labelled flow-based data set is CTU-13. It was especially created for training botnet detection algorithms. CTU-13 contains a variety of botnet traffic mixed with background traffic coming from a real network [16].

Shiravi et al. [50] introduce a dynamic data set approach based on profiles containing abstract representations of events and network behaviour. This allows to generate reproducible, modified and extended data sets for better comparison of different IDS. We use unidirectional flows whereas [50] contains bidirectional ones. Converting flows might constitute a viable approach, but would require some effort for re-labelling.

In conclusion, the validation of IDS through data sets in general seems to be difficult due to few publicly available data sets. Some of the most widely used data sets are outdated [17, 48, 57] or only contain malicious data which complicates the attempt for comprehensive validation. Since we use unidirectional flows some of the presented data sets would only be applicable after conversion and re-labelling of the data [50]. Some data sets are only for specific attack scenarios [16] and other promising approaches are not publicly available [62, 66].

4.2   Data Set Emulation

Małowidzki et al. [31] define a list of characteristics for a good data set. A good data set should contain recent, realistic and labelled data. It should be rich containing all the typical attacks met in the wild as well as be correct regarding operating cycles in enterprises, e.g. working hours. We try to meet these the requirements listed in [31] in our approach using *OpenStack*.

OpenStack is an open source software platform which allows the creation of virtual networks and virtual machines. This platform provides certain advantages when generating flow-based data sets. A test environment can be easily scaled by using virtual machines and therefore allows to generate data sets of any size. Furthermore, one has full control over the environment including the control of the network traffic. This ensures the correct capturing and labelling of truly clean flow-based data sets which do not contain any harmful scenarios. Conversely, it is also possible to include a host as an attacker and clearly label the data generated from the attacker as malicious. A data set acquired from a real enterprise network can never be labelled with the same quality. However, it is of upmost importance to emulate the network activity as authentic as possible to generate viable data sets which is difficult for synthetic data. To reach that goal, we emulate a sample small company network with different subnets containing various servers and simulated clients and record generated network traffic in unidirectional *NetFlow* format.

4.2.1 Test Environment

Our sample network currently comprises three subnets, namely a management, a developer and a server subnet. Each subnet contains several servers or clients. A schematic overview of the network is shown in Fig. 7.

The servers provide different services like printing on a network printer, email, file sharing and hosting websites. The clients, which are windows and linux machines, emulate user activity through various Python scripts. All three subnets are connected through a virtual router. *Open vSwitch* is used to monitor the traffic in the virtual network and to capture it in *NetFlow* format.

As already mentioned, *OpenStack* easily enables scaling of the desired test environment. Our setup scripts contain the automatic update of the newly created hosts as well as the installation of tools required to emulate normal user activity and setting up an automatic startup of the said activities. Using this approach, new hosts can easily be added to the test environment if needed.

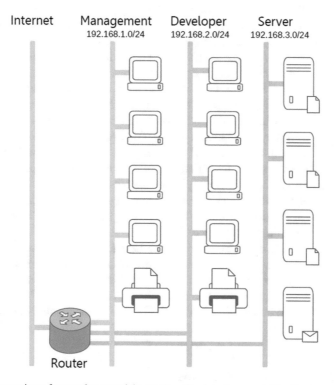

Fig. 7 An overview of a sample network in our *OpenStack* environment. The *Router* separates the internal network from the internet and acts as firewall. The internal network structure with three subnets containing several clients and servers is shown on the *right*

4.2.2   Generation of Normal Data

Several aspects have to be considered when emulating normal user behaviour through scripts. Ideally, scripts should:

1. run on all established operating systems
2. include all computerized activities of typical employees
3. be free of periodic activities
4. consider different working methods
5. consider working hours and idling of the employees

to be as close to real user behaviour as possible. Using this list, we have set up a kind of user model within the scripts which reflects the real behaviour of the users. The fact that we use the platform-independent language Python resolves the first issue above.

To be realistic, emulated clients have to simulate a broad list of activities. Those activities include all quintessential computerized activities of a typical employee like browsing the web for work-related matters (e.g. companies websites or doing research via search engines), private browsing (e.g. facebook, certain newsgroups or blogs), file sharing, writing emails or printing on a network printer. Also, it is important to ensure a varying length of files regarding the latter two activities and varying types and numbers of attachments in the activity of sending emails.

To emulate user behaviour as realistically as possible it is vital to not simply repeat given activities periodically but rather involve randomness to some extend. Conversely, the activities should not be totally random but reflect different working methods of the employees. It should be noted that some employees are more likely to write a great amount of emails but rarely use a file sharing service. Configuration files can be altered to model the different characteristica of each type of employee.

During a typical work day, employees are not permanently using "their" assigned computer. Meetings, offline work or coffee breaks should also be taken into account. The last point of the list is similar. The activities typically focus on the employees working hours and drop during lunch break or in the evening.

It is to be noted that while we do our best to satisfy the requirements for generating as realistic data as possible, we do not claim to capture emulating viable behaviour. In essence, one needs to ensure making the simulation as realistic as possible by modelling time and size distributions. While those distributions are not implemented yet, in this setup new ideas for improving the emulation can be easily integrated and tested.

4.2.3   Generation of Malicious Data

In addition to clean non-threatening flow-based data, malicious data behaviour is needed in a fully labelled data set. This problem is solved by inserting one or more hosts deployed as attacker. Subsequently, it is easy to label malicious flows using the structure of the *NetFlow* data as well as additional knowledge like the *Source*

IP Address of the attacker and the exact timestamp of the attack. Again, Python scripts are used to create malicious behaviour like *DoS* attacks, *Port Scans*, or *SSH brute force* attacks. If malicious data is inserted via scripts, it is always possible to include new attacks by simply writing new scripts. Thus, up-to-date data sets can be generated at any time.

5   Related Work

Work on network based anomaly detection methods for intrusion and insider threat detection can be separated into packet-based and flow-based anomaly detection. A comprehensive review of both methods is given in Bhuyan et al. [3]. A recent survey of data mining and machine learning methods for cyber security intrusion detection is published by Buczak and Guven [4]. Further, Weller-Fahy et al. [61] published an overview of similarity measures which are used for anomaly based network intrusion detection.

Since the proposed approach is based on flow-based data, the following review does not consider packet-based methods. We categorize flow-based anomaly detection methods into (I) treating each flow separately, (II) aggregating all flows over time windows and (III) aggregating flows of single hosts over time windows.

Category I

Winter et al. [63] propose a flow-based anomaly detection method of category (I). The authors use an One-Class SVM and train their system with malicious flows instead of benign flows since data mining methods are better at finding similarities than outliers. For learning the One-Class SVM, the honeypot data set of [53] is used. During the evaluation phase, each flow within the class is considered as malicious and each outlier is considered as normal behaviour. Another approach of this category is proposed by Tran et al. [58]. The basis of their system is a block-based neural network (BBNN) integrated within an FPGA. They extract four attributes (*Packets*, *Bytes*, *Duration* and *Flags*) from each flow as input for their IDS. The authors compared their system against SVM and Naive Bayes classifier and outperformed them in an experimental evaluation. Najafabadi et al. [35] use four different classification algorithms for *SSH Brute Force* detection. The authors selected eight attributes from the flow-based data and were able to detect the attacks. The detection of *RUDY* attacks using classification algorithms is studied by Najafabadi et al. [36]. *RUDY* is an *application layer DoS attack* which generates much less traffic than traditional *DoS* attacks. The authors use the enriched flow-based SANTA dataset [62] for evaluation. The flows in this data set contain additional attributes which are calculated based on full packet captures.

Category II

Approaches from category (II) aggregate all flows within a certain time window. Wagner et al. [60] developed a special kernel function for anomaly detection. The authors divide the data stream in equally sized time windows and consider each time window as a data point for their kernel. The kernel function takes information about the *Source (Destination) IP Address* and the transferred *Bytes* of all flows within the time window. The authors integrate their kernel function in an One-Class SVM and evaluate their approach in the context of an internet service provider (ISP). An entropy based anomaly detection approach is presented in [38]. Here, the authors divide the data stream in 5 min intervals and calculate for each interval seven different distributions considering flow-header attributes and behavioural attributes. Based on these distributions, entropy values are calculated which are used for anomaly detection.

Category III

Approaches from category (III) use more preprocessing algorithms in the data mining workflow and do not work directly on flow-based data. These approaches aggregate for each host the flows over a time window and calculate new attributes based on these aggregations. BClus [16] uses this approach for behavioural-based botnet detection. At first, they divide the flow data stream in time windows. Then, flows are aggregated by *Source IP Address* for each time window. For each aggregation, new attributes (e.g. amount of unique destination IP addresses contacted by this *Source IP Address*) are calculated and used for further analysis. The authors evaluate their botnet detection approach using the CTU-13 Malware data set. Another representative of this category is proposed by Najafabadi et al. [34]. The authors aggregate all *NetFlows* with the same *Source IP Address*, *Destination IP Address* and *Destination Port* in 5 min intervals. Based on these aggregations, new attributes are calculated like the *average transmitted bytes* or the *standard deviation of the transmitted bytes*. Then, Najafabadi et al. [34] train different classifiers and use them for the detection of *SSH Brute Force* attacks.

Besides these three categories we want to also mention the Apache Spot[5] framework. Apache Spot is an open source framework for analysing packet- and flow based network traffic on Hadoop. This framework allows to use machine learning algorithms for identifying malicious network traffic. Another network based anomaly detection system is proposed by Rehak et al. [42, 43]. Their system *Camnep* uses various anomaly detectors and combine their results to decide if the network traffic is normal or malicious. The individual filters use direct attributes from *NetFlow* and additional context attributes. For calculating these attributes, the anomaly detectors can access all flows of a 5 min window.

---

[5]http://open-network-insight.org/.

Our proposed approach neither handles each flow separately (category I) nor simply aggregates all flows within a time window (category II). Instead, it follows the approach of the third category and collects all flows for each host within a time window. However, in contrast to the third category, the proposed approach generates more than one data point for each collection. The generation of multiple data points for each collection allows us to calculate more adapted data points which describe the network, service and user behaviour of the hosts. Further, we do not try to recognize all attack types with a single classifier like [60] or [63]. Instead, we analyse the calculated data points from different views and develop for each view a separate detection engine.

6 Summary and Future Work

Company data needs to be protected against unauthorized access. Attempts to gain such unauthorized access may originate from outside the company network, but may also be traced back to insiders. In this contribution, we report on ongoing work that aims to assist domain experts by highlighting significant incidents. To that end, we develop the Coburg Utility Framework (CUF), a toolset to support the analysis using data mining methods. CUF is based on a pipes-and-filter architecture and contains various machine learning algorithms, visualization tools, preprocessing algorithms and evaluation algorithms. Due to its architecture, CUF is very flexible and can be easily extended or adapted.

While our earlier work focused on offline analysis of data from company networks, we currently work on extending our approach to online analysis of data streams. In particular, our approach builds upon flow-based data since this allows to reduce the volume of data, bypasses the problem of encrypted payloads and leads to less privacy concerns compared to packet-based data. Further, we include additional domain knowledge to support otherwise uniformed analysis methods. In particular, our approach includes analyses from three different points of view: the first perspective is concerned with the general network behaviour of hosts, the second perspective focuses on the usage of services, and the third perspective concentrates on user behaviour.

Evaluation of anomaly-based intrusion and insider threat detection approaches presupposes test data that are labelled in terms of normal or malicious behaviour. As it turns out, such labelled test data are hard to obtain since such data sets are rarely available for public use. Likewise, real data from company networks cannot be used easily due to the lack of reliable labels. Therefore, we devised and set up a virtual environment for generating flow-based network data. To that end, an *OpenStack* environment may emulate a flexible configuration of servers and clients. Client behaviour within that virtual network is generated through randomized Python scripts. This allows to record real flow-based traffic with typical user activities and further to simulate attacks like *Port Scans*, *DoS* or *SSH Brute Force*.

Future activities of our research are directed towards refining the flow-based analysis approach and provide appropriate visualization tools for data streams, e.g. by extending well-known visualization approaches such as parallel coordinates to data streams. In addition, the simulation environment needs to be expanded allowing to generate even more realistic data sets as a basis to validate and refine our anomaly-based intrusion and insider threat detection approach more thoroughly.

Acknowledgements This work is funded by the Bavarian Ministry for Economic affairs through the WISENT project (grant no. IUK 452/002).

References

1. Aggarwal, C.C., Han, J., Wang, J., Yu, P.S.: A framework for clustering evolving data streams. In: International Conference on very large data bases (VLDB), pp. 81–92. Morgan Kaufmann (2003)
2. Agrawal, R., Gehrke, J., Gunopulos, D., Raghavan, P.: Automatic subspace clustering of high dimensional data for data mining applications. In: International Conference on Management of Data, pp. 94–105. ACM Press (1998)
3. Bhuyan, M.H., Bhattacharyya, D.K., Kalita, J.K.: Network anomaly detection: Methods, systems and tools. IEEE Communications Surveys & Tutorials **16**(1), 303–336 (2014)
4. Buczak, A.L., Guven, E.: A survey of data mining and machine learning methods for cyber security intrusion detection. IEEE Communications Surveys & Tutorials **18**(2), 1153–1176 (2016)
5. Caliński, T., Harabasz, J.: A dendrite method for cluster analysis. Communications in Statistics-theory and Methods **3**(1), 1–27 (1974)
6. Cao, F., Ester, M., Qian, W., Zhou, A.: Density-based clustering over an evolving data stream with noise. In: SIAM International Conference on Data Minning (SDM), vol. 6, pp. 328–339. Society for Industrial and Applied Mathematics (2006)
7. Chae, H.s., Jo, B.o., Choi, S.H., Park, T.: Feature selection for intrusion detection using NSL-KDD. Recent Advances in Computer Science pp. 978–960 (2015)
8. Chen, E.Y.: Detecting DoS attacks on SIP systems. In: IEEE Workshop on VoIP Management and Security, 2006., pp. 53–58. IEEE (2006)
9. Chou, C.H., Su, M.C., Lai, E.: A new cluster validity measure and its application to image compression. Pattern Analysis and Applications **7**(2), 205–220 (2004)
10. Claise, B.: Cisco systems netflow services export version 9. RFC 3954 (2004)
11. Claise, B.: Specification of the ip flow information export (IPFIX) protocol for the exchange of ip traffic flow information. RFC 5101 (2008)
12. Datti, R., Verma, B.: B.: Feature reduction for intrusion detection using linear discriminant analysis. International Journal on Engineering Science and Technology **1**(2) (2010)
13. Davies, D.L., Bouldin, D.W.: A cluster separation measure. IEEE transactions on pattern analysis and machine intelligence **1**(2), 224–227 (1979)
14. Depren, O., Topallar, M., Anarim, E., Ciliz, M.K.: An intelligent intrusion detection system (IDS) for anomaly and misuse detection in computer networks. Expert systems with Applications **29**(4), 713–722 (2005)
15. Fayyad, U.M., Irani, K.B.: Multi-interval discretization of continuous-valued attributes for classification learning. In: International Joint Conference on Artificial Intelligence (IJCAI), pp. 1022–1029. Morgan Kaufmann (1993)
16. Garcia, S., Grill, M., Stiborek, J., Zunino, A.: An empirical comparison of botnet detection methods. Computers & Security **45**, 100–123 (2014)

17. Gharibian, F., Ghorbani, A.A.: Comparative study of supervised machine learning techniques for intrusion detection. In: Annual Conference on Communication Networks and Services Research (CNSR'07), pp. 350–358. IEEE (2007)
18. Giacinto, G., Perdisci, R., Del Rio, M., Roli, F.: Intrusion detection in computer networks by a modular ensemble of one-class classifiers. Information Fusion **9**(1), 69–82 (2008)
19. Goseva-Popstojanova, K., Anastasovski, G., Pantev, R.: Using multiclass machine learning methods to classify malicious behaviors aimed at web systems. In: International Symposium on Software Reliability Engineering, pp. 81–90. IEEE (2012)
20. Guha, S., Rastogi, R., Shim, K.: Rock: A robust clustering algorithm for categorical attributes. In: International Conference on Data Engineering, pp. 512–521. IEEE (1999)
21. Hassani, M., Seidl, T.: Internal clustering evaluation of data streams. In: Trends and Applications in Knowledge Discovery and Data Mining, pp. 198–209. Springer (2015)
22. Hellemons, L., Hendriks, L., Hofstede, R., Sperotto, A., Sadre, R., Pras, A.: SSHCure: a flow-based SSH intrusion detection system. In: IFIP International Conference on Autonomous Infrastructure, Management and Security, pp. 86–97. Springer (2012)
23. John, W., Dusi, M., Claffy, K.C.: Estimating routing symmetry on single links by passive flow measurements. In: International Wireless Communications and Mobile Computing Conference, pp. 473–478. ACM (2010)
24. Jung, J., Paxson, V., Berger, A.W., Balakrishnan, H.: Fast portscan detection using sequential hypothesis testing. In: IEEE Symposium on Security and Privacy, pp. 211–225. IEEE (2004)
25. Kang, D.K., Fuller, D., Honavar, V.: Learning classifiers for misuse and anomaly detection using a bag of system calls representation. In: Annual IEEE SMC Information Assurance Workshop, pp. 118–125. IEEE (2005)
26. Kendall, K.: A database of computer attacks for the evaluation of intrusion detection systems. Tech. rep., DTIC Document (1999)
27. Landes, D., Otto, F., Schumann, S., Schlottke, F.: Identifying suspicious activities in company networks through data mining and visualization. In: P. Rausch, A.F. Sheta, A. Ayesh (eds.) Business Intelligence and Performance Management, pp. 75–90. Springer (2013)
28. Lee, C.H.: A hellinger-based discretization method for numeric attributes in classification learning. Knowledge-Based Systems **20**(4), 419–425 (2007)
29. Lin, J., Lin, H.: A density-based clustering over evolving heterogeneous data stream. In: ISECS International Colloquium on Computing, Communication, Control, and Management, vol. 4, pp. 275–277. IEEE (2009)
30. Liu, Q., Dong, G.: CPCQ: Contrast pattern based clustering quality index for categorical data. Pattern Recognition **45**(4), 1739–1748 (2012)
31. Małowidzki, M., Berezinski, P., Mazur, M.: Network intrusion detection: Half a kingdom for a good dataset. In: NATO STO SAS-139 Workshop, Portugal (2015)
32. Moore, A.W., Zuev, D.: Internet traffic classification using bayesian analysis techniques. In: ACM SIGMETRICS International Conference on Measurement and Modeling of Computer Systems, pp. 50–60. ACM, New York, USA (2005)
33. Murtagh, F., Contreras, P.: Algorithms for hierarchical clustering: an overview. Wiley Interdisciplinary Reviews: Data Mining and Knowledge Discovery **2**(1), 86–97 (2012)
34. Najafabadi, M.M., Khoshgoftaar, T.M., Calvert, C., Kemp, C.: Detection of SSH brute force attacks using aggregated netflow data. In: International Conference on Machine Learning and Applications (ICMLA), pp. 283–288. IEEE (2015)
35. Najafabadi, M.M., Khoshgoftaar, T.M., Kemp, C., Seliya, N., Zuech, R.: Machine learning for detecting brute force attacks at the network level. In: International Conference on Bioinformatics and Bioengineering (BIBE), pp. 379–385. IEEE (2014)
36. Najafabadi, M.M., Khoshgoftaar, T.M., Napolitano, A., Wheelus, C.: Rudy attack: Detection at the network level and its important features. In: International Florida Artificial Intelligence Research Society Conference (FLAIRS), pp. 288–293 (2016)
37. Nguyen, T.T., Armitage, G.: A survey of techniques for internet traffic classification using machine learning. IEEE Communications Surveys & Tutorials **10**(4), 56–76 (2008)

38. Nychis, G., Sekar, V., Andersen, D.G., Kim, H., Zhang, H.: An empirical evaluation of entropy-based traffic anomaly detection. In: ACM SIGCOMM Conference on Internet measurement, pp. 151–156. ACM (2008)

39. Phaal, P., Panchen, S., McKee, N.: InMon Corporation's sFlow: A Method for Monitoring Traffic in Switched and Routed Networks. RFC 3176 (2001)

40. Pramana, M.I.W., Purwanto, Y., Suratman, F.Y.: DDoS detection using modified k-means clustering with chain initialization over landmark window. In: International Conference on Control, Electronics, Renewable Energy and Communications (ICCEREC), pp. 7–11 (2015)

41. Rampure, V., Tiwari, A.: A rough set based feature selection on KDD CUP 99 data set. International Journal of Database Theory and Application **8**(1), 149–156 (2015)

42. Rehák, M., Pechoucek, M., Bartos, K., Grill, M., Celeda, P., Krmicek, V.: Camnep: An intrusion detection system for high-speed networks. Progress in Informatics **5**(5), 65–74 (2008)

43. Rehák, M., Pechoucek, M., Grill, M., Stiborek, J., Bartoš, K., Celeda, P.: Adaptive multiagent system for network traffic monitoring. IEEE Intelligent Systems **24**(3), 16–25 (2009)

44. Ring, M., Otto, F., Becker, M., Niebler, T., Landes, D., Hotho, A.: Condist: A context-driven categorical distance measure. In: European Conference on Machine Learning and Knowledge Discovery in Databases, pp. 251–266. Springer (2015)

45. Ring, M., Wunderlich, S., Grüdl, D., Landes, D., Hotho, A.: Flow-based benchmark data sets for intrusion detection. In: Proceedings of the 16th European Conference on Cyber Warfare and Security (ECCWS). ACPI (2017, to appear)

46. Rossi, D., Valenti, S.: Fine-grained traffic classification with netflow data. In: International wireless communications and mobile computing conference, pp. 479–483. ACM (2010)

47. Rostamipour, M., Sadeghiyan, B.: An architecture for host-based intrusion detection systems using fuzzy logic. Journal of Network and Information Security **2**(2) (2015)

48. Shah, V.M., Agarwal, A.: Reliable alert fusion of multiple intrusion detection systems. International Journal of Network Security **19**(2), 182–192 (2017)

49. Shearer, C.: The CRISP-DM model: the new blueprint for data mining. Journal of data warehousing **5**(4), 13–22 (2000)

50. Shiravi, A., Shiravi, H., Tavallaee, M., Ghorbani, A.A.: Toward developing a systematic approach to generate benchmark datasets for intrusion detection. Computers & Security **31**(3), 357–374 (2012)

51. Skoudis, E., Liston, T.: Counter Hack Reloaded: A Step-by-step Guide to Computer Attacks and Effective Defenses. Prentice Hall Series in Computer Networking and Distributed Systems. Prentice Hall Professional Technical Reference (2006)

52. Sommer, R., Paxson, V.: Outside the closed world: On using machine learning for network intrusion detection. In: IEEE Symposium on Security and Privacy, pp. 305–316. IEEE (2010)

53. Sperotto, A., Sadre, R., Van Vliet, F., Pras, A.: A labeled data set for flow-based intrusion detection. In: IP Operations and Management, pp. 39–50. Springer (2009)

54. Sridharan, A., Ye, T., Bhattacharyya, S.: Connectionless port scan detection on the backbone. In: IEEE International Performance Computing and Communications Conference, pp. 10–pp. IEEE (2006)

55. Staniford, S., Hoagland, J.A., McAlerney, J.M.: Practical automated detection of stealthy portscans. Journal of Computer Security **10**(1-2), 105–136 (2002)

56. Tan, P.N., Steinbach, M., Kumar, V.: Introduction to Data Mining, (First Edition). Addison-Wesley Longman Publishing Co., Inc., Boston, MA, USA (2005)

57. Tavallaee, M., Bagheri, E., Lu, W., Ghorbani, A.A.: A detailed analysis of the KDD CUP 99 data set. In: IEEE Symposium on Computational Intelligence for Security and Defense Applications, pp. 1–6 (2009)

58. Tran, Q.A., Jiang, F., Hu, J.: A real-time netflow-based intrusion detection system with improved BBNN and high-frequency field programmable gate arrays. In: International Conference on Trust, Security and Privacy in Computing and Communications, pp. 201–208. IEEE (2012)

59. Valenti, S., Rossi, D., Dainotti, A., Pescapè, A., Finamore, A., Mellia, M.: Reviewing traffic classification. In: Data Traffic Monitoring and Analysis, pp. 123–147. Springer (2013)

60. Wagner, C., François, J., Engel, T., et al.: Machine learning approach for ip-flow record anomaly detection. In: International Conference on Research in Networking, pp. 28–39. Springer (2011)
61. Weller-Fahy, D.J., Borghetti, B.J., Sodemann, A.A.: A survey of distance and similarity measures used within network intrusion anomaly detection. IEEE Communications Surveys & Tutorials **17**(1), 70–91 (2015)
62. Wheelus, C., Khoshgoftaar, T.M., Zuech, R., Najafabadi, M.M.: A session based approach for aggregating network traffic data - the santa dataset. In: International Conference on Bioinformatics and Bioengineering (BIBE), pp. 369–378. IEEE (2014)
63. Winter, P., Hermann, E., Zeilinger, M.: Inductive intrusion detection in flow-based network data using one-class support vector machines. In: International Conference on New Technologies, Mobility and Security (NTMS), pp. 1–5. IEEE (2011)
64. Yang, C., Zhou, J.: Hclustream: A novel approach for clustering evolving heterogeneous data stream. In: International Conference on Data Mining-Workshops (ICDMW'06), pp. 682–688. IEEE (2006)
65. Zander, S., Nguyen, T., Armitage, G.: Automated traffic classification and application identification using machine learning. In: The IEEE Conference on Local Computer Networks 30th Anniversary (LCN'05) l, pp. 250–257. IEEE (2005)
66. Zuech, R., Khoshgoftaar, T.M., Seliya, N., Najafabadi, M.M., Kemp, C.: A new intrusion detection benchmarking system. In: International Florida Artificial Intelligence Research Society Conference (FLAIRS), pp. 252–256. AAAI Press (2015)

Human-Machine Decision Support Systems for Insider Threat Detection

Philip A. Legg

Abstract Insider threats are recognised to be quite possibly the most damaging attacks that an organisation could experience. Those on the inside, who have privileged access and knowledge, are already in a position of great responsibility for contributing towards the security and operations of the organisation. Should an individual choose to exploit this privilege, perhaps due to disgruntlement or external coercion from a competitor, then the potential impact to the organisation can be extremely damaging. There are many proposals of using machine learning and anomaly detection techniques as a means of automated decision-making about which insiders are acting in a suspicious or malicious manner, as a form of large scale data analytics. However, it is well recognised that this poses many challenges, for example, how do we capture an accurate representation of normality to assess insiders against, within a dynamic and ever-changing organisation? More recently, there has been interest in how visual analytics can be incorporated with machine-based approaches, to alleviate the data analytics challenges of anomaly detection and to support human reasoning through visual interactive interfaces. Furthermore, by combining visual analytics and active machine learning, there is potential capability for the analysts to impart their domain expert knowledge back to the system, so as to iteratively improve the machine-based decisions based on the human analyst preferences. With this combined human-machine approach to decision-making about potential threats, the system can begin to more accurately capture human rationale for the decision process, and reduce the false positives that are flagged by the system. In this work, I reflect on the challenges of insider threat detection, and look to how human-machine decision support systems can offer solutions towards this.

P.A. Legg (✉)
Department of Computer Science and Creative Technologies,
University of the West of England, Bristol, UK
e-mail: phil.legg@uwe.ac.uk

© Springer International Publishing AG 2017
I. Palomares et al. (eds.), *Data Analytics and Decision Support for Cybersecurity*,
Data Analytics, DOI 10.1007/978-3-319-59439-2_2

33

1   Introduction

It is often said that for any organisation, "employees are the greatest asset, and yet also the greatest threat". The challenge of how to address this *insider threat* is one that is of increasing concern for many organisations. In particular, as our modern world is rapidly evolving, so to are the ways in that we conduct business and manage organisations, and so to are the ways in that those who choose to attack can do so, and succeed. In recent times there have been many high profile cases, including Edward Snowden [1], Bradley Manning [2], and Robert Hanssen [3]. According to the 2011 CyberSecurity Watch Survey [4], whilst 58% of cyber-attacks on organisations are attributed to outside threats, 21% of attacks are initiated by their own employees or trusted third parties. In the Kroll 2012 Global Fraud Survey [5], they report that 60% of frauds are committed by insiders, up from 55% in the previous year. Likewise, the 2012 Cybercrime report by PwC [6] states that the most serious fraud cases were committed by insiders. Of course, in all of these cases, these figures may not truly reflect the severity of the problem given that there are most likely many more that are either not detected, or not reported publicly. To define what is an 'insider', it is often agreed that this is somebody who compared to an outsider, has some level of knowledge and some level of access in relation to an organisation. Whilst employees are often considered to be the main focal point as insiders, by this definition there may be many others, such as contractors, stakeholders, former employees, and management, who could also be considered as insiders.

Insider threat research has attracted a significant amount of attention in the literature due to the severity of the problem within many organisations. Back in 2000, early workshops on insider threat highlighted the many different research challenges surrounding the topic [7]. Since then, there have been a number of proposals to address these challenges. For example, Greitzer et al. [8] discuss strategies for combating the insider-threat problem, including raising staff awareness and more effective methods for identifying potential risks. In their work, they define an insider to be an individual who currently, or at one time, was authorised to access an organisation's information system, data, or network. Likewise, they refer to an insider threat as a harmful act that trusted insiders might carry out, such as causing harm to an organisation, or an unauthorised act that benefits the individual. Carnegie Mellon University has conducted much foundational work surrounding the insider-threat problem as part of their CERT (Computer Emergency Response Team), resulting in over 700 case-studies that detail technical, behavioural, and organisational details of insider crimes [9]. They define a malicious insider to be a current or former employee, contractor, or other business partner who has or had authorized access to an organisation's network, system, or data and intentionally exceeded or misused that access in a manner that negatively affected the confidentiality, integrity, or availability of the organisation's information or information systems. Spitzner [10] discusses early research on insider-threat detection using honeypots (decoy machines that may lure an attack). However, as security awareness increases, those choosing to commit insider attacks are finding more subtle methods to cause harm or defraud their organisations, and so there is a need for more sophisticated prevention and detection.

In this chapter, I discuss and reflect on my recent research that addresses the issues that surround insider threat detection. Some of this work has been previously published in various journals and conference venues. The contribution that this chapter serves is to bring together previous work on developing automated machine-based detection tools, and to reconsider the problem of insider threat detection with regards to how the human and the machine can work in tandem to identify malicious activity. Neither the human alone, nor the machine alone, is sufficient to address the problem in a satisfactory manner.

2 Related Works

There are a variety of published works on the topic of insider threat detection that range from theoretical frameworks for representing the problem domain, through to practical implementations of detection systems. As a research area, it is multi-disciplinary in nature, including computational design of detection algorithms, human behavioural modelling, business operations management, and ethical and legal implications of insider surveillance.

2.1 Models for Understanding the Problem of Insider Threat

Legg et al. propose a conceptual model that can help organisations to begin thinking about how to detect and prevent insider attacks [11]. The model is based on a tiered approach that relates real-world activity, measurement of the activity, and hypotheses about the current threat. The model is designed to capture a broad range of attributes related to insider activity that could be characterised by some means. The tiered approach aims to address how multiple attributes from the real-world tier can contribute towards the collection of measurements that may prove useful for forming hypotheses (e.g., heavy workload, working late, and a developing disagreement with higher management, could result in a possible threat of sabotage). Nurse et al. [12] also propose a framework, this time for characterising insider threat activity. The framework is designed to help an analyst identify the various traits that surround insider threats, including the precipitating events that then motivate an attacker, and the identification of resources and assets that may be exploited as part of an attack. By considering these attributes, analysts may be able to ensure a full and comprehensive security coverage in their organisation.

Maybury et al. [13] developed a taxonomy for the analysis and detection of insider threat that goes beyond only cyber actions, to also incorporate such measures as physical access, violations, finances and social activity. Similarly, Colwill [14] examines the human factors surrounding insider threat in the context of a large telecommunications organisation, remarking that greater education and awareness of the problem is required, whilst Greitzer et al. [15] focus on incorporating inferred psychological factors into a modelling framework. The work by Brdiczka et al. [16] combine such psychological profiling with structural anomaly detection, to develop an architecture for insider-threat detection that demonstrates much potential for solving the problem.

In terms of measuring behaviours that may indicate a threat, Roy et al. [17] propose a series of metrics that could be used based on technical and behavioural observations. Schultz [18] presents a framework for prediction and detection of insider attacks. He acknowledges that no single behavioural clue is sufficient to detect insider threat, and so suggest using a mathematical representation of multiple indicators, each with a weighted contribution. Althebyan and Panda [19] present a model for insider-threat prediction based on the insider's knowledge and the dependency of objects within the organisation. In the work of Sasaki [20], a trigger event is used to identify a change of behaviour, that impel an insider to act in a particular way (for instance, if the organisation announce an inspection, an insider threat may begin deleting their tracks and other data records).

Bishop et al. [21] discuss the insider-threat problem, and note that the term insider threat is ill-defined, and rightly recognise that there should be a degree of "insiderness" rather than a simple binary classification of insider threat or not. They propose the Attribute-Based Group Access Control (ABGAC) model, as a generalisation of role-based access control, and show its application to three case studies [22]: embezzlement, social engineering, and password alteration. Other work such as Doss and Tejay [23] propose a model for insider-threat detection that consists of four stages: monitoring, threat assessment, insider evaluation and remediation. Liu et al. [24] propose a multilevel framework called SIDD (Sensitive Information Dissemination Detection) that incorporates network-level application identification, content signature generation and detection, and covert communication detection. More recently, Bishop et al. [25] extend their work to examine process modelling as a means for detecting insider attacks.

2.2 Approaches for Detecting Insider Threat

Agrafiotis et al. [26] explore the sequential nature of behavioural analysis for insider threat detection. The sequence of events is a critical aspect of analysis, since a single event in isolation may not be deemed as a threat, and yet in conjunction with other events, this may have much greater significance. As an example, an employee who is accessing sensitive company records would be of more concern if they had recently been in contact with a rival organisation, compared to an employee who may be acting as part of their job role requirement. They extend the work on sequential analysis in [27], where this scheme is then applied to characterise a variety of insider threat case studies that have been collated by the Carnegie Mellon University CERT.

Elmrabit et al. [28] study the categories and approaches of insider threat. They categorise different types of insider attack (e.g., sabotage, fraud, IP theft) against the CIA security principles (confidentiality, integrity, availability), and also against human factors (motive, opportunity, capability). They discuss a variety of tools in the context of insider threat detection, such as intrusion detection systems, honeytokens, access control systems, and security information and event management systems. They also highlight the importance of psychological prediction models, and security education and awareness, both of which are required by organisations in order to tackle the insider threat problem effectively. It is clear that technical measures alone are not sufficient, and that 'security as a culture' should be practiced by organisations wishing to address this issue successfully.

Parveen et al. [29] use stream mining and graph mining to detect insider activity in large volumes of streaming data, based on ensemble-based methods, unsupervised learning and graph-based anomaly detection. Building on this, Parveen and Thuraisingham [30] propose an incremental learning algorithm for insider threat detection that is based on maintaining repetitive sequences of events. They use trace files collected from real users of the Unix C shell, however this public dataset is relatively dated now. Buford et al. [31] use situation-aware multi-agent systems as part of a distributed architecture for insider threat detection. Garfinkel et al. [32] propose tools for media forensics, as means to detecting insider threat behaviour.

Eldardiry et al. [33] also propose a system for insider threat detection based on feature extraction from user activities, although they do not consider role-based assessments as part of their system. Senator et al. [34] propose to combine structural and semantic information on user behaviour to develop a real-world detection system. They use a real corporate database, gather as part of the Anomaly Detection at Multiple Scales (ADAMS) program, however due to confidentiality they can not disclose the full details and so it is difficult to compare against the work.

McGough et al. [35] propose a beneficial software system for insider threat detection based on anomaly detection of a user profile and their job role profile. Their approach also aims to incorporate human resources information, for which they describe a five states of happiness approach to assess the likelihood that a user may pose a threat. Nguyen and Reiher [36] propose a detection tool for insider threat that monitors system call activity for unusual or suspicious behaviour. Maloof and Stephens [37] propose a detection tool for when insiders violate need-to-know restrictions that are in place within the organisation. Okolica et al. [38] use Probabilistic Latent Semantic Indexing with Users to determine employee interests, which are used to form social graphs that can highlight insiders.

2.3 *Insider Threat Visualization*

With regards to insider threat visualization, the technical report by Harris [39] discusses some of the issues related to visualizing insider threat activity. Nance and Marty [40] propose using bipartite graphs to identify and visualize insider threat activity where the nodes in the graph represent two distinct groups, such as user nodes and activity nodes, and the edges represent that a particular user has performed a particular activity. This approach is best suited for comparative analysis once a small group of users and activities have been identified, as scalability issues would soon arise in most real-world analysis tasks. Stoffel et al. [41] propose a visual analytics application for identifying correlations between different networked devices, based on time-series anomaly detection and similarity models. They focus primarily at the network traffic level, and so they do not currently consider other attributes related to insider threat such as file storage systems and USB connected devices. Kintzel et al. [42] use scalable glyph-based visualization using a clock metaphor to present an overview of the activity over time of thousands of hosts on a network. Zhao et al. [43] looked at anomaly detection for social media data and presented their visualization tool FluxFlow. Again, they make use of the clock metaphor as part of their visualization, which they combine with scaled circular glyphs to represent anomalous data points. Walton et al. [44] proposed QCATs (Multiple Queries with Conditional Attributes) as a technique for understanding and visualizing conditional probabilities in the context of anomaly detection.

2.4 Summary of Related Works

From the literature it becomes clear to see that the topic of insider threat has been extensively studied from a variety of viewpoints. A number of models have been put forward for how one could observe and detect signs that relate to whether an insider is posing a threat, or has indeed already attacked. Likewise, a number of detection techniques have been proposed. However, it is difficult to assess their true value when some only consider a sub-set of activities, or do not provide validation in a real-world context. In the following sections, I discuss work that has been conducted in recent years on insider threat detection by colleagues and myself. In particular, I address both machine-based and human-based approaches for decision-making on the current threat posed by an individual. As part of this, I also describe the real-world validation study of the machine-driven decision process that was performed, and an active learning approach for combining human-machine decision-making using visual analytic tools. These contributions set the work apart from the wider body of research that exists on insider threat detection, by supporting both human and machine in the process of identifying malicious insiders.

3 Automated Detection of Insider Threats

The process of detecting insiders that pose suspicious or malicious activity is a complex challenge. Given the large volume of data that may exist about all users activity within an organisation, human methods alone will not prove scalable.

Instead, there is a need for the machine to make a well-informed decision about the threat posed by an individual, based on their observed activity, and this differs from what is deemed as normal behaviour.

3.1   Automated Detection Using User and Role-Based Profile Assessment

In the paper by Legg et al. [45], "Automated Insider Threat Detection System using User and Role-based Profile Assessment", an insider threat detection system is proposed that is capable of identifying anomalous activity of users, in comparison to their previous activity and in comparison to their peers . The detection tool is based upon the underlying principles of the conceptual model proposed in [11]. The paper demonstrates the detection tool using publicly-available insider threat datasets provided by Carnegie Mellon University CERT, along with ten synthetic scenarios that were generated by an independent team within the Oxford Cyber Security group. In the work, the requirements of the detection system are given that:

– The system should be able to determine a score for each user that relates to the threat that they currently pose.
– The system should be able to deal with various forms insider threat, including sabotage, intellectual property theft, and data fraud.
– The system should also be able to deal with unknown cases of insider threat, whereby the threat is deemed to be an anomaly for that user and for that role.
– The system should assess the threat that an individual poses based on how this behaviour deviates from both their own previous behaviour, and the behaviour exhibited by those in a similar job role.

The system comprises of five key components: data input streams, user and role-based profiling, feature extraction, threat assessment, and classification of threat. From the data streams that were available for the CMU-CERT scenarios, and for those developed by the Oxford team, the data typically represented the actions of 1000 employees over the period of 12 months, with data that captured login and logout information for PC workstations, USB device insertion and removal, file access, http access, and e-mail communications. Each user also has an assigned job role (e.g., technician, receptionist, or director), where those in a similar role are expected to share some commonality in their behaviour. The first stage of the system is to connect to the available data streams, and to receive data from each stream in the correct time sequence as given by the timestamp of each activity.

As data is received, this is utilised to populate a profile that represents each individual user, as well as a combined profile that represents a single role. The profiles are constructed in a consistent hierarchical fashion, that denotes the devices that have been accessed by the user, the actions performed on each of these devices,

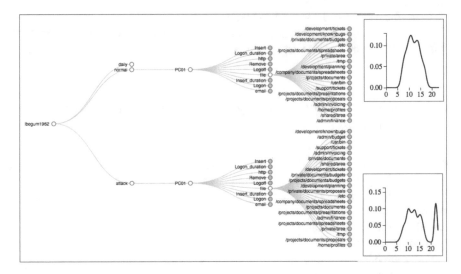

Fig. 1 Tree-structured profiles of user and role behaviours. The root node is the user ID, followed by sub branches for 'daily', 'normal', and 'attack' observations. The next level down shows devices used, then activities performed, and finally, attributes for those activities. The probability distribution for normal hourly usage is given in the *top-right*, and the distribution for the detected attack is given in the *bottom-right*. Here it can be seen that the user has accessed a new set of file resources late at night

and the attributes associated with these actions. At each of these nodes in the profile, a time-series is constructed that denotes the occurrence of observations on a 24-h period. Figure 1 shows an interactive tree view of an individual user's profile.

Once the system has computed the current daily profile for each user and for each role, the system can then extract features from the profile. Since the profile structure is consistent and well-defined, it means that comparisons between users, roles, or time steps can be easily made. In particular, the feature sets consists of three main categories: the user's daily observations, comparisons between the user's daily activity and their previous activity, and comparisons between the user's daily activity and the previous activity of their role. The full set of features that are computed for each user is provided in [45]. These features include a variety of measurements that can be derived from the profiles, such as *New device for user*, *New attribute for activity for device for role*, *Hourly usage count for activity*, *USB duration for user*, and *Earliest logon time for user*. This set of features intends to be widely applicable for most organisations, although of course, there may be more bespoke features that are relevant for specific organisations that could also be incorporated. To perform the threat assessment, the system aims to identify variance between related features that may be indicative of a particularly anomaly. This is performed using Principal Component Analysis (PCA) [46]. PCA performs a projection of the features into lower dimensional space based on the amount of variance exhibited by each feature. From the user profiles, an $n \times m$ matrix is constructed for each

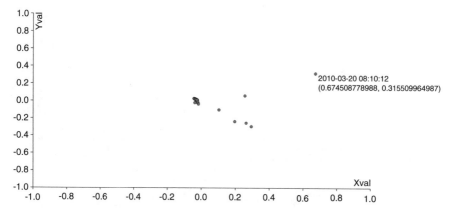

Fig. 2 Example of using PCA for assessing deviation in user activity. Each point represents a single user for a single day (observation instance). Here, only a single user over time is shown to preserve clarity. The majority of points form a cluster at the centre of the plot. There are five observations that begin to move away from the general cluster. At the far right is a point observed on the 20th March 2010, that exhibits the most deviation in the user's behaviour

user, where n is the total number of sessions (or days) being considered, and m is the number of features that have been obtained from the profile. The bottom row of the matrix represents the current daily observation, with the remainder of the matrix being all previous observation features. Essentially, this process reduces an n-dimensional dataset to $n - 1$ dimensionality, based on the vector of greatest variance through the data. By performing this successively, we can reduce to 2 or 3 dimensions. Similar instances would be expected to group together, whilst instances that exhibit significant variation would appear far from other points in the space, where each point represents a single user on a single day. The system performs PCA using a variety of different feature combination that relate to a particular area of concern (e.g., web activity). Figure 2 shows the PCA decomposition for a detected insider threat. It can be seen that the most-part of activity clusters towards the centre, however over time, there are activities that diverge from this cluster, that represent daily observations where the user has performed significantly different. By considering the Euclidean distance of points from the centroid of the cluster, or from the point given by the role average, a measure of anomaly or deviation can be obtained for a given observation.

The score for each anomaly metric can then be analysed, for each user, for each day (e.g., *file_anomaly*, *total_anomaly*, *role_anomaly*). A parallel co-ordinate plot is used (Fig. 3), where each polyline shows a single user for a single day, against the various anomaly metrics (where each axis is a separate anomaly metric). In the example shown in Fig. 3, there is an observation that appears separate on the *any_anomaly* metric (this relates to activity that has been observed on **any** device— rather than just **this** device that it may have been observed on). By brushing the axis, the analyst can filter the view to show only this result. This reveals activity performed by a particular user of interest, who was found to be the malicious insider

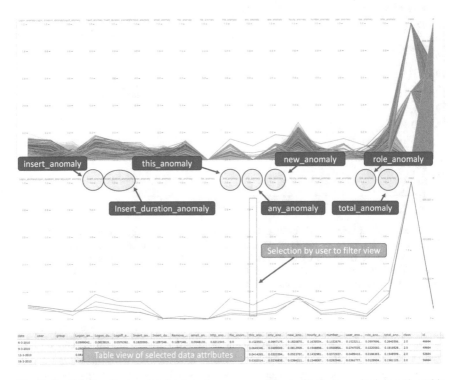

Fig. 3 Parallel Coordinates view to show the corresponding profile features. An interactive table below the parallel co-ordinates view shows a numerical data view of the profile features that have been selected. Here, a particular user scores significantly higher than other users on one metric. Interactive brushing allows this to be examined in further detail

in the set of 1000 employees, who was accessing systems and using a USB storage device early in the morning. This was found to be different from the other users in the role of 'Director', who did not use USB storage devices, and very rarely used systems at this time of day.

This approach was found to be successful for the test scenarios from CMU-CERT and from Oxford. Unlike other supervised machine learning techniques, this approach requires no labelling of instances, making it easier to be deployed quickly and effectively within an organisation. Given the variety of ways that a user may exhibit activity that could be deemed a threat, classifying instances may be quite difficult in any case. Classification also assumes that future instances of a particular threat will closely match the currently-observed case, which may not be the case (e.g., exfiltration of data could be performed in a variety of ways). The challenge with the proposed approach is ensuring that the available data streams can capture the occurrence of the activity that is required to identify the threat. It also requires that the feature extraction supports all possible attack vectors that could be imagined that relates to the available data. Whilst a comprehensive set of features are provided, organisations may well find that they wish to incorporate additional features, or refine the existing features. By adopting an extensive approach for obtaining features, modification and creation of new features can be achieved with minimal reconfiguration.

3.2 Validation of Decision Support in Insider Threat Systems

Whilst the previous section describes the detection tool in detail, perhaps the biggest challenge with developing insider threat tools is actually validating their performance in the real-world. Previously, synthetic data scenarios were used for developing and testing the tool. The paper by Agrafiotis et al. [47], "Validating an insider threat detection system: A real scenario perspective" extends this to report on the deployment of the detection tool in a real organisation.

The head of security for the particular organisation in question (not disclosed for confidentiality purposes) indicated that there had recently been an incident, which meant that there was a known insider that the system could be trialled against. From discussions with the head of security, the detection system was modified to account for three particular areas of interest: File-access logs, Patent DB interactions, and Directory DB interactions. Compared to the previous work of [45], the real-world organisation also presented scalability challenges. Here, file access logs provided more than 750,000 data entries per day, compared to approximately 20,000 in the synthetic examples. However, by only considering authenticated data entries resulted in a significant reduction in the amount of data from 750,000 to 44,000 entries per day. This was deemed as appropriate by the head of security, since this then provided user details, whereas the unauthenticated attempts were simply denied access to the system. They use five anomaly metrics (which are anonymised in their paper due to non-disclosure agreements), based on combinations of the features derived from the user activity profile.

For testing the system, they deployed the detection system over two different time periods (1 September to 31 October, and 1 December to 31 December), accounting for 16,000 employees. The December period contained no known cases, and served as a training period for establishing a baseline of normal activity to compare against. The period in September and October contained one known case of insider threat activity. When testing on this dataset, a number of false positives were generated as either medium or high alert, for 4129 individuals. However, on closer inspection, what the authors actually found was that the system produced approximately 0.5 alerts per employee per day. Yet, for one particular user, they generated 12 alerts in a single day. Sure enough, this particular user was the insider. Given the nature of a multi-national organisation, working times are likely to change significantly, and it is recognised by the head of security that users do not conform to strict working patterns on a regular basis. However, the fact that the system is capable of identifying the repeat occurrence of alerts for a user shows the strong potential of this system. Further work aims to consider how combinations of alerts across multiple days can be accumulated to better separate this particular individual from the other alerts that were generated. Nevertheless, the importance of this study is crucial for the continued development of insider threat detection tools, and demonstrates a real-world validation of how a system can be deployed in a large complex organisation.

4 Visual Analytics of Insider Threat Detection

The previous section looked at machine-based approaches for detecting insider threat activity from a large group of users. The capable of the machine to make such well-informed decisions is, by large, limited by the data that is available to the system, and how the system can understand and make sense of features that are derived from the user profiles. This section looks to explore how the human can utilise this knowledge that the machine generates, to further improve the decision-making process. Realistically, the disciplinary action of an insider would not be enforced until a security analyst and management have gathered the facts and can confidently identify that the user is a threat. Therefore, the machine-based approach serves to reduce the search space that the human analyst needs to consider, and then the human can explore this further, to understand *why* the machine may have arrived at such a decision, and whether the human agrees or disagrees with this decision.

4.1 Supporting Human Reasoning using Interactive Visual Analytics

In the paper by Legg [48], "Visualizing the Insider Threat: Challenges and tools for identifying malicious user activity", it is shown how visualization can be utilised to better support the decision-making process of the detection tool. The system makes use of a visual analytics dashboard, supported by a variety of linked views including a interactive PCA (iPCA) view (as originally proposed by Jeong et al. [49]). The proposed dashboard, shown in Fig. 4, allows for overview summary statistics to be viewed, based on selection of time, users, and job roles. The iPCA view shows the measurement features on a parallel coordinates plot, and a scatter plot that represents the 2-dimensional PCA. In particular, what this offers is the ability to observe how the PCA space relates back to the original feature space. By dragging points in the scatter plot, a temporary black polyline is displayed on the parallel co-ordinates that shows the inverse PCA for the new dragged position, giving an interactive indication of how the 2-dimensional space maps to the original feature space. For the analyst, this can be particularly helpful to strengthen their reasoning for a particular hypothesis, such as for understanding what a particular cluster of points may be indicative of. The tool also features an activity view, where activities are plotted by time in a radial view (Fig. 5). This can be particularly useful for examining the raw

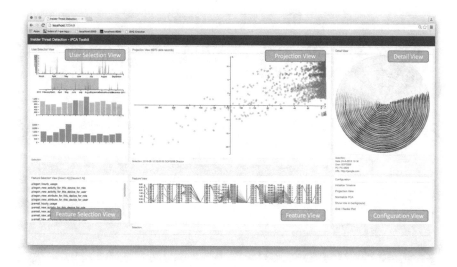

Fig. 4 Layout of the visual analytics dashboard. The dashboard consists of four visualization views: User Selection, Projection, Detail, and Feature. The dashboard also has two supporting views for feature selection and configuration

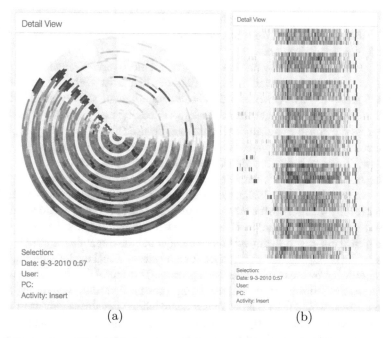

Fig. 5 Two variants of the detail view for exploring user activity, using (**a**) a circular plot (where time maps to angle and day maps to the radius), or (**b**) a rectangular grid plot (where time maps to the x-axis and day maps to the y-axis). *Colour* denotes the observed activity, and the selection pane provides detail of attributes. The role profile can be shown by the translucent *coloured* segments

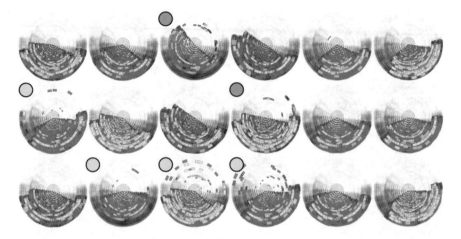

Fig. 6 Assessment of 18 different user profiles within the same job role. Of the profiles, six profiles exhibit activity that occurs outside of the typical time period (marked by a *circle top-left* of profile). Two of the users also use USB devices (marked by a *blue circle*) during this non-typical time period, which may be of potential interest to the analyst. This view provides a compact and comparable overview of similar users

activity for days where there is significant deviation. Again, this links with the PCA view, so that when a user hovers on a point, the corresponding ring in the radial view is highlighted, and visa versa.

The detail view also forms the basis for a role overview mode, where the analyst can inspect the detail view of all users that exist within the same role. Figure 6 shows 18 users, where red indicates login/logout activity, blue indicates USB insertion/removal, green indicates e-mail activity, and yellow indicates web activity. As previously, a translucent background is used to represent the role profile, so that comparisons can be made between how the user compares against this. From this view, it can be seen that six of the users access resources outside of typical working hours for that particular role (marked by a circle top-left of profile), and two of these are making use of USB devices during these non-typical hours (marked by a blue circle). By visualizing the activity by means of this overview, it allows analysts to gain a clearer understanding of how the other users in the role perform, which can help further support their decision-making about the threat that is posed.

Visual Analytics provide an interface for analysts to visual explore the analytical results produced by the machine decision process. The ability to utilise the machine-based detection allows for an initial filtering process that alleviates the workload for the analyst, whilst the visual analytics approach then enables analysts to obtain a richer understanding of why the machine has given a particular result, without obscuring the data should the analyst decide that further analytics of additional data is required to fulfil their own decision process.

4.2   Active Learning for Insider Threat Detection

The visual analytics dashboard is a powerful interface that links the user to intuitive visual representations of the underlying data. Through interaction, the analyst can explore and delve deeper into this data, to support the development of their hypotheses on the intentions of an insider who may pose a threat to the organisation.

By exploiting the concept of a visual analytics loop [50], the user interaction can be utilised to inform the system, based on the new knowledge that they have obtained from viewing the current result. From a machine learning viewpoint, this is akin to the human providing online training labels, based on instances of particular interest. This concept of training on a small sample of key instances, as determined by the current result of the system (rather than providing a complete training set like in supervised learning) is referred to as *active learning* [51].

In the paper by Legg et al. [52], "Caught in the act of an insider attack: detection and assessment of insider threat", an active learning approach is proposed for refining the configuration of the detection system. Figure 7 shows the approach for introducing active learning into a visual analytics tool. As seen previously, a parallel co-ordinates plot is used to depict each user for each day. The plot can be configured to show a historical time windows (e.g., the last 30 days). The left-side shows a user alert list, where minor and severe alerts are show as orange and red respectively, and the date and user are given as the text label. If the analyst clicks on an alert, a tree-structured profile is displayed (Fig. 1) that allows them to explore deeper into *why* a particular observation has been flagged. In the tree profile, all previously-acceptable activity is shown under the *normal* node, whilst the current attack is shown under the *attack* node. In this example, it appears that the user has accessed a set of files late at night that they would not typically work with, hence why they have been flagged up in this case.

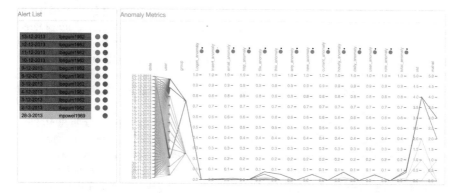

Fig. 7 Detection system as a result of active learning. The analyst has rejected the alert on *mpowell1969* (shown by the removal of the accept option). This reconfigures the detection system to downgrade the anomaly associated with this result—in this case *insert_anomaly*—which can be observed by the *circular dials* by each anomaly metric. In addition to the alert list, the parallel co-ordinates can be set to present only the 'last 30 days', which provides a clear view of the detected insider *lbegum1962*

For the active learning component, the key here is that each label also has an accept or reject option (shown by the green and red circles to the right of the label). The user does not necessarily have to provide this information, however if they do, then the system is able to incorporate this knowledge based into the decision-making process. This is done by taking a weighted contribution from each feature, so that if a rejected result scores highly on a particular feature, then this feature can be down-weighted for this particular user, or role, or entire group, for a particular period of time. In this fashion, the burden of false positives can be alleviated for the analysts.

5 Future Directions for Insider Threat Detection

It should be apparent by now that there is substantial interest in the area of insider threat detection as a research discipline. Yet despite many proposed solutions, the problem continues to persist in many organisations. So why is this? Part of the challenge is security awareness. Many organisations are simply ill-equipped to gather and analyse such activity data. Others may choose not to invest in security until it is too late. Part of the challenge here is also how to transition the work from academic research into industrial practice. A number of spin-out companies are beginning to come about from insider-threat research, for which may begin to address this problem. So, what is it that organisations can be doing to protect themselves?

Perhaps the key element is for organisations to identify their most precious assets, and identifying features to represent these. Machine Learning routines can only form useful insight if working with appropriate data that is representative of the problem domain. A system is unlikely to detect that a user is about to steal many sensitive records if it knows nothing about a user's access to such records. Therefore, identifying which activity features an organisation is most concerned about is a vital step in any application of insider threat detection. It is also vital that appropriate visual analytic tools are in place for assessing the results of automated detection routines, so that the analyst can fully understand the reasoning behind why a particular individual has been flagged up as suspicious. Without this, humans are merely taking a machine's word for whether a individual should be disciplined. Given the severe consequence of false accusations, it is vital that the analyst has full confidence in a given decision.

Another emerging area of interest in combating the insider threat is analysing text communications. This raises many ethical and privacy concerns, although in a corporate environment it could be argued that this is a requirement of the role (e.g., employees working in national security would be expected to abide by such regulations). One proposal to provide analytics on textual data without exposing privacy concerns is to perform online linguistics analysis that can then be used to characterise the communication, rather than the raw text alone. In [53], the Linguistic Enquiry Word Count (LIWC) tool was used as a means of characterising psychological traits through use of language. The LIWC tool essentially provides dictionaries that relate particular words (or parts of words), to independent features (e.g., love, friend, hate, self). There has been much work in the psychology domain of relating LIWC features to OCEAN characteristics (Openness, Conscientiousness, Extroversion, Agreeableness, Neuroticism) [54], and also the Dark Triad (Narcissism, Machiavellianism, Psychopathy) [55]. A visual analytics dashboard was developed for analysing the communications of multiple users against these features, that can be used to identify when there is significant change in a user's communication, and how this could imply a change in their psychological characteristics. Whilst this initial study demonstrated potential in this area, there is much work that remains to be done in how feasible a solution this can provide.

Another consider to make is who should be responsible for decision making in insider threat—human or machine? Given the severity of disciplinary action, it could be argued that a human should always need to intervene to inspect the result to ensure that this is valid before any disciplinary or legal action is taken. Then there is the issue of at what stage does the system intervene—should analysts operate as proactive or reactive? In a proactive environment, systems may attempt to predict the likelihood that a user will become an attacker, rather than a reactive environment that is detecting an already-conducted attack. Again, ethical concerns are raised such as whether a user would have conducted the attack if the system had not intervened at that time? Lab-based experimentation on such scenarios can only take us so far in understanding these problems, and the concept that an employee can be disciplined for an action that they are yet to perform is very much seen as the work of science fiction (much like that of *Minority Report*). However, what is required is the collaboration and cooperation between organisations, and also with academia, to continue to experiment and continue to develop tools that can alleviate and support the demands of the analyst in their decision-making process.

6 Summary

In this chapter, I have considered the scope of insider threat detection, and how systems can be developed to enable human-machine decision support such that well-informed decisions can be made about insider threats. Whilst a wide range of work exists on the topic, there is still much work to be done to combat the insider threat. How can detection systems be provided with accurate and complete data? How can detection systems extend beyond 'cyber' data sources, to build a more complete representation of the organisation? How should the psychology of insiders be accounted for, to understand their motives and intentions in their normal practice, and understand how these may change and why? Then there are the ethical concerns that need to be addressed—if employees are being monitored, how will this affect staff morale? Will they simply find alternative ways to circumvent protective measures?

The research shows much potential in being able to combat this problem, however, it also reveals the importance of the human aspects of security. As stated earlier, "employees are the greatest asset, and yet also the greatest threat", and it is fair that this has never been so true as it is in our modern society today. Technology is enhancing how society operate, and yet it is also providing new means for disgruntled insiders to attack. At the same time, insiders acting in physical space are also becoming more creative in their planning. This begins to illustrate how the boundaries between online and offline worlds are beginning to blur, and cyber is just another factor in the much larger challenge of organisation security. Yet, with continued efforts in the development of new security technologies, we can better support the decision-making process between man and machine, to combat the challenge of insider threat detection.

Acknowledgements Many thanks to my colleagues from Oxford Cyber Security, Dr. Ioannis Agrafiotis, Dr. Jassim Happa, Dr. Jason Nurse, Dr. Oliver Buckley (now with Cranfield University), Professor Michael Goldsmith, and Professor Sadie Creese, with whom my early work on insider threat detection was carried out with.

References

1. BBC News. Profile: Edward Snowden, 2013. http://www.bbc.co.uk/news/world-us-canada-22837100.
2. Guardian. Bradley manning prosecutors say soldier 'leaked sensitive information', 2013. http://www.guardian.co.uk/world/2013/jun/11/bradley-manning-wikileaks-trial-prosecution.
3. FBI. Robert Philip Hanssen Espionage Case, 2001. http://www.fbi.gov/about-us/history/famous-cases/robert-hanssen.
4. CSO Magazine, CERT Program (Carnegie Mellon University) and Deloitte. CyberSecurity Watch Survey: Organizations Need More Skilled Cyber Professionals To Stay Secure, 2011. http://www.sei.cmu.edu/newsitems/cybersecurity_watch_survey_2011.cfm.
5. Kroll and Economist Intelligence Unit. Annual Global Fraud Survey. 2011/2012, 2012.
6. PricewaterhouseCoopers LLP. Cybercrime: Protecting against the growing threat - Events and Trends, 2012.
7. R. Anderson, T. Bozek, T. Longstaff, W. Meitzler, M. Skroch, and K. Van Wyk. Research on mitigating the insider threat to information systems. In *Proceedings of the Insider Workshop, Arlington, Virginia, USA*. RAND, August 2000.
8. F. L. Greitzer, A. P. Moore, D. M. Cappelli, D. H. Andrews, L. A. Carroll, and T. D. Hull. Combating the insider cyber threat. *Security & Privacy, IEEE*, 6(1):61–64, 2007.
9. D. M. Cappelli, A. P. Moore, and R. F. Trzeciak. *The CERT Guide to Insider Threats: How to Prevent, Detect, and Respond to Information Technology Crimes*. Addison-Wesley Professional, 1st edition, 2012.
10. L. Spitzner. Honeypots: catching the insider threat. In *Proc. of the 19th IEEE Computer Security Applications Conference (ACSAC'03), Las Vegas, Nevada, USA*, pages 170–179. IEEE, December 2003.
11. P. A. Legg, N. Moffat, J. R. C. Nurse, J. Happa, I. Agrafiotis, M. Goldsmith, and S. Creese. Towards a conceptual model and reasoning structure for insider threat detection. *Journal of Wireless Mobile Networks, Ubiquitous Computing, and Dependable Applications*, 4(4):20–37, 2013.

12. J. R. C. Nurse, O. Buckley, P. A. Legg, M. Goldsmith, S. Creese, G. R. T. Wright, and M. Whitty. Understanding insider threat: A framework for characterising attacks. In *Security and Privacy Workshops (SPW), 2014 IEEE*, pages 214–228, May 2014.
13. M. Maybury, P. Chase, B. Cheikes, D. Brackney, S. Matzner, T. Hetherington, B. Wood, C. Sibley, J. Marin, T. Longstaff, L. Spitzner, J. Haile, J. Copeland, and S. Lewandowski. Analysis and detection of malicious insiders. In *Proc. of the International Conference on Intelligence Analysis, McLean, Viginia, USA*. MITRE, May 2005.
14. C. Colwill. Human factors in information security: The insider threat who can you trust these days? *Information Security Technical Report*, 14(4):186–196, 2009.
15. F. L. Greitzer and R. E. Hohimer. Modeling human behavior to anticipate insider attacks. *Journal of Strategic Security*, 4(2):25–48, 2011.
16. O. Brdiczka, J. Liu, B. Price, J. Shen, A. Patil, R. Chow, E. Bart, and N. Ducheneaut. Proactive insider threat detection through graph learning and psychological context. In *Proc. of the IEEE Symposium on Security and Privacy Workshops (SPW'12), San Francisco, California, USA*, pages 142–149. IEEE, May 2012.
17. K. R. Sarkar. Assessing insider threats to information security using technical, behavioural and organisational measures. *Information Security Technical Report*, 15(3):112–133, 2010.
18. E. E. Schultz. A framework for understanding and predicting insider attacks. *Computers and Security*, 21(6):526–531, 2002.
19. Q. Althebyan and B. Panda. A knowledge-base model for insider threat prediction. In *Proc. of the IEEE Information Assurance and Security Workshop (IAW'07), West Point, New York, USA*, pages 239–246. IEEE, June 2007.
20. T. Sasaki. A framework for detecting insider threats using psychological triggers. *Journal of Wireless Mobile Networks, Ubiquitous Computing, and Dependable Applications*, 3(1/2): 99–119, 2012.
21. M. Bishop, S. Engle, S. Peisert, S. Whalen, and C. Gates. We have met the enemy and he is us. In *Proc. of the 2008 workshop on New security paradigms (NSPW'08), Lake Tahoe, California, USA*, pages 1–12. ACM, September 2008.
22. M. Bishop, S. Engle, S. Peisert, S. Whalen, and C. Gates. Case studies of an insider framework. In *Proc. of the 42nd Hawaii International Conference on System Sciences (HICSS'09), Waikoloa, Big Island, Hawaii, USA*, pages 1–10. IEEE, January 2009.
23. G. Doss and G. Tejay. Developing insider attack detection model: a grounded approach. In *Proc. of the IEEE International conference on Intelligence and security informatics (ISI'09), Richardson, Texas, USA*, pages 107–112. IEEE, June 2009.
24. Y. Liu, C. Corbett, K. Chiang, R. Archibald, B. Mukherjee, and D. Ghosal. SIDD: A framework for detecting sensitive data exfiltration by an insider attack. In *System Sciences, 2009. HICSS '09. 42nd Hawaii International Conference on*, pages 1–10, Jan 2009.
25. M. Bishop, B. Simidchieva, H. Conboy, H. Phan, L. Osterweil, L. Clarke, G. Avrunin, and S. Peisert. Insider threat detection by process analysis. In *IEEE Security and Privacy Workshops (SPW)*. IEEE, 2014.
26. I. Agrafiotis, P. A. Legg, M. Goldsmith, and S. Creese. Towards a user and role-based sequential behavioural analysis tool for insider threat detection. *J. Internet Serv. Inf. Secur.(JISIS)*, 4(4):127–137, 2014.
27. I. Agrafiotis, J. R. C. Nurse, O. Buckley, P. A. Legg, S. Creese, and M. Goldsmith. Identifying attack patterns for insider threat detection. *Computer Fraud & Security*, 2015(7):9 – 17, 2015.
28. N. Elmrabit, S. H. Yang, and L. Yang. Insider threats in information security categories and approaches. In *Automation and Computing (ICAC), 2015 21st International Conference on*, pages 1–6, Sept 2015.
29. P. Parveen, J. Evans, Bhavani Thuraisingham, K.W. Hamlen, and L. Khan. Insider threat detection using stream mining and graph mining. In *Privacy, security, risk and trust (passat), 2011 ieee third international conference on and 2011 ieee third international conference on social computing (socialcom)*, pages 1102–1110, Oct 2011.
30. P. Parveen and B. Thuraisingham. Unsupervised incremental sequence learning for insider threat detection. In *Intelligence and Security Informatics (ISI), 2012 IEEE International Conference on*, pages 141–143, June 2012.

31. J. F. Buford, L. Lewis, and G. Jakobson. Insider threat detection using situation-aware mas. In *Proc. of the 11th International Conference on Information Fusion*, pages 1–8, 2008.
32. S. L. Garfinkel, N. Beebe, L. Liu, and M. Maasberg. Detecting threatening insiders with lightweight media forensics. In *Technologies for Homeland Security (HST), 2013 IEEE International Conference on*, pages 86–92, Nov 2013.
33. H. Eldardiry, E. Bart, Juan Liu, J. Hanley, B. Price, and O. Brdiczka. Multi-domain information fusion for insider threat detection. In *Security and Privacy Workshops (SPW), 2013 IEEE*, pages 45–51, May 2013.
34. T. E. Senator, H. G. Goldberg, A. Memory, W. T. Young, B. Rees, R. Pierce, D. Huang, M. Reardon, D. A. Bader, E. Chow, et al. Detecting insider threats in a real corporate database of computer usage activity. In *Proceedings of the 19th ACM SIGKDD international conference on Knowledge discovery and data mining*, pages 1393–1401. ACM, 2013.
35. A. S. McGough, D. Wall, J. Brennan, G. Theodoropoulos, E. Ruck-Keene, B. Arief, C. Gamble, J. Fitzgerald, A. van Moorsel, and S. Alwis. Insider threats: Identifying anomalous human behaviour in heterogeneous systems using beneficial intelligent software (ben-ware). In *Proceedings of the 7th ACM CCS International Workshop on Managing Insider Security Threats*, MIST '15, pages 1–12, New York, NY, USA, 2015. ACM.
36. N. Nguyen and P. Reiher. Detecting insider threats by monitoring system call activity. In *Proceedings of the 2003 IEEE Workshop on Information Assurance*, 2003.
37. M. A. Maloof and G. D. Stephens. elicit: A system for detecting insiders who violate need-to-know. In Christopher Kruegel, Richard Lippmann, and Andrew Clark, editors, *Recent Advances in Intrusion Detection*, volume 4637 of *Lecture Notes in Computer Science*, pages 146–166. Springer Berlin Heidelberg, 2007.
38. J. S. Okolica, G. L. Peterson, and R. F. Mills. Using plsi-u to detect insider threats by datamining e-mail. *International Journal of Security and Networks*, 3(2):114–121, 2008.
39. M. Harris. Visualizing insider activity and uncovering insider threats. Technical report, 2015.
40. K. Nance and R. Marty. Identifying and visualizing the malicious insider threat using bipartite graphs. In *System Sciences (HICSS), 2011 44th Hawaii International Conference on*, pages 1–9, Jan 2011.
41. F. Stoffel, F. Fischer, and D. Keim. Finding anomalies in time-series using visual correlation for interactive root cause analysis. In *Proceedings of the Tenth Workshop on Visualization for Cyber Security*, VizSec '13, pages 65–72, New York, NY, USA, 2013. ACM.
42. C. Kintzel, J. Fuchs, and F. Mansmann. Monitoring large ip spaces with clockview. In *Proceedings of the 8th International Symposium on Visualization for Cyber Security*, VizSec '11, pages 2:1–2:10, New York, NY, USA, 2011. ACM.
43. J. Zhao, N. Cao, Z. Wen, Y. Song, Y. Lin, and C. Collins. Fluxflow: Visual analysis of anomalous information spreading on social media. *Visualization and Computer Graphics, IEEE Transactions on*, 20(12):1773–1782, Dec 2014.
44. S. Walton, E. Maguire, and M. Chen. Multiple queries with conditional attributes (QCATs) for anomaly detection and visualization. In *Proceedings of the Eleventh Workshop on Visualization for Cyber Security*, VizSec '14, pages 17–24, New York, NY, USA, 2014. ACM.
45. P. A. Legg, O. Buckley, M. Goldsmith, and S. Creese. Automated insider threat detection system using user and role-based profile assessment. *IEEE Systems Journal*, PP(99):1–10, 2015.
46. I. Jolliffe. *Principal component analysis*. Wiley Online Library, 2005.
47. I. Agrafiotis, A. Erola, J. Happa, M. Goldsmith, and S. Creese. Validating an insider threat detection system: A real scenario perspective. In *2016 IEEE Security and Privacy Workshops (SPW)*, pages 286–295, May 2016.
48. P. A. Legg. Visualizing the insider threat: challenges and tools for identifying malicious user activity. In *Visualization for Cyber Security (VizSec), 2015 IEEE Symposium on*, pages 1–7, Oct 2015.
49. D. H. Jeong, C. Ziemkiewicz, B. Fisher, W. Ribarsky, and R. Chang. ipca: An interactive system for pca-based visual analytics. In *Proceedings of the 11th Eurographics / IEEE - VGTC Conference on Visualization*, EuroVis'09, pages 767–774, Chichester, UK, 2009. The Eurographs Association; John Wiley & Sons, Ltd.

50. P. A. Legg, D. H. S. Chung, M. L. Parry, R. Bown, M. W. Jones, I. W. Griffiths, and M. Chen. Transformation of an uncertain video search pipeline to a sketch-based visual analytics loop. *Visualization and Computer Graphics, IEEE Transactions on*, 19(12):2109–2118, Dec 2013.

51. B. Settles. Active learning. *Synthesis Lectures on Artificial Intelligence and Machine Learning*, 6(1):1–114, 2012.

52. P. A. Legg, O. Buckley, M. Goldsmith, and S. Creese. Caught in the act of an insider attack: detection and assessment of insider threat. In *Technologies for Homeland Security (HST), 2015 IEEE International Symposium on*, pages 1–6, April 2015.

53. P. A. Legg, O. Buckley, M. Goldsmith, and S. Creese. Visual analytics of e-mail sociolinguistics for user behavioural analysis. *Journal of Internet Services and Information Security (JISIS)*, 4(4):1–13, 2014.

54. J. S. Wiggings. *The five factor model of personality: Theoretical perspectives*. Guilford Press, 1996.

55. D. L. Paulhus and K. M. Williams. The dark triad of personality: Narcissism, machiavellianism, and psychopathy. *Journal of research in personality*, 36(6):556–563, 2002.

Detecting Malicious Collusion Between Mobile Software Applications: The Android™ Case

Irina Măriuca Asăvoae, Jorge Blasco, Thomas M. Chen,
Harsha Kumara Kalutarage, Igor Muttik, Hoang Nga Nguyen,
Markus Roggenbach, and Siraj Ahmed Shaikh

Abstract Malware has been a major problem in desktop computing for decades. With the recent trend towards mobile computing, malware is moving rapidly to smartphone platforms. "Total mobile malware has grown 151% over the past year", according to McAfee®'s quarterly treat report in September 2016. By design, Android™ is "open" to download apps from different sources. Its security depends on restricting apps by combining digital signatures, sandboxing, and permissions. Unfortunately, these restrictions can be bypassed, without the user noticing, by colluding apps for which combined permissions allow them to carry out attacks. In this chapter we report on recent and ongoing research results from our ACID

I.M. Asăvoae
INRIA, Paris, France
e-mail: irina-mariuca.asavoae@inria.fr

J. Blasco (✉)
Information Security Group, Royal Holloway University of London, Egham, UK
e-mail: jorge.blascoalis@rhul.ac.uk

T.M. Chen
School of Mathematics, Computer Science & Engineering, City University of London, London, UK
e-mail: tom.chen.1@city.ac.uk

H.K. Kalutarage
Centre for Secure Information Technologies, Queen's University of Belfast, Belfast, UK
e-mail: h.kalutarage@qub.ac.uk

I. Muttik
Cyber Curio LLP, Berkhamsted, UK
e-mail: igor.muttik@cybercurio.com

H.N. Nguyen • S.A. Shaikh
Centre for Mobility and Transport, Coventry University, Coventry, UK
e-mail: hoang.nguyen@coventry.ac.uk; siraj.shaikh@coventry.ac.uk

M. Roggenbach
Department of Computer Science, Swansea University, Swansea, UK
e-mail: m.roggenbach@swansea.ac.uk

© Springer International Publishing AG 2017
I. Palomares et al. (eds.), *Data Analytics and Decision Support for Cybersecurity*,
Data Analytics, DOI 10.1007/978-3-319-59439-2_3

project which suggest a number of reliable means to detect collusion, tackling the aforementioned problems. We present our conceptual work on the topic of collusion and discuss a number of automated tools arising from it.

1 Introduction

One of the most fundamental principles in computer security is ensuring effective isolation of software. Programs written by different independent software vendors (ISVs) have to be properly separated to avoid any accidental data flows as well as all deliberate data leaks. Strictly speaking, even subroutines of a program need to be properly isolated and some computer systems attempt that too via, for example, protection of stack frames and memory tagging. This isolation principle helps ensure both reliability of software (by limiting the effect of design flaws, insecure design, bugs, etc.) as well as protect from outright malicious code (malicious data-leaking libraries, exploiting vulnerabilities via injecting shell-code, etc.)

The era of personal computing slightly diminished the requirement for isolation. It was believed that PCs—being single-user devices—will be OK with all software running at the same privilege. First personal computers had no hardware and software support for software isolation. However, reliability and privacy demanded a better solution so these primitive OSes were replaced by multiple descendants of Windows NT®, Unix® and Linux®. Requirements for better isolation also drove special hardware features—examples include Intel® SGX enclaves and ARM® TrustZone®.

In the cloud environments (like Docker®, Amazon EC2, Microsoft Azure®, etc.) which execute software from different sources and operate on data belonging to different entities, guaranteed isolation becomes even more important because any cross-container data leaks (deliberate or accidental) may be devastating. Communications across cloud containers have to be covert because no explicit APIs is provided. The problem of covert communication between programs running in time-sharing computer systems was first discussed as early as in 1970s [36].

The situation is quite different in mass-market operating systems for mobile devices such as smart phones—there is no need for covert channels at all. While corresponding operating systems (Symbian®, MeeGo®, iOS®, Android™, Tizen®, etc.) were designed with the isolation principle in mind, the requirement for openness led to the ability of software in the device to communicate in many different ways. Android™ OS is a perfect example of such a hybrid design—apps run in sandboxes but they have documented means of sending and receiving messages to/from each other; they can also create shared objects and files. These inter-app communication mechanisms are handy but, unfortunately, also make it possible to carry out harmful actions in a collaborative fashion.

Extreme commonality of Android™ as well as rapid growth of malicious and privacy-leaking apps made it a perfect target for our team to look for colluding behaviours. Authors of malicious software would be interested in flying under the radar for as long as possible. Unscrupulous advertisers could also benefit from hiding privacy-invading functionality in multiple apps. These reasons led us to believe that Android™ may be one of the first targets for collusion attacks. We also realised that security practitioners who analyse threats for Android™ desperately need tools which would help them uncover colluding apps. Such apps may be outright malicious or they may be unwanted programs which often do aggressive advertising coupled with disregard for users' privacy (like those which would use users' contacts to expand their advertising further). Having a popular OS which allowed (and to some extent even provides support to) colluding apps was a major risk.

Before we started there were no tools or established methods to uncover these attacks: discovering such behaviours is very tricky—two or more mobile apps, when analysed independently, may not appear to be malicious. However, together they could become harmful by exchanging information with one another. Multi-app threats such as these were considered theoretical for some years, but as part of this research we discovered colluding code embedded in many Android™ apps in the wild [48]. Our goal was to find effective methods of detecting colluding apps in Android™ [6–8, 11–13, 37]. This would potentially pave a way for spotting collusions in many other environments that implement software sandboxing, from other mobile operating systems to virtual machines in server farms.

1.1 Background

Malware has been a major problem in desktop computing for decades. With the recent trend towards mobile computing, malware is moving rapidly to mobile platforms. "Total mobile malware has grown 151% over the past year", according to McAfee®'s quarterly threat report from September 2016. Criminals are clearly motivated by the opportunity—the number of smartphones in use is predicted to grow from 2.6 billion in 2016 to 6.1 billion in 2020, predominantly Android™, with more than 10 billion apps downloaded to date. Smartphones pose a particular security risk because they hold personal details (accounts, locations, contacts, photos) and have potential capabilities for eavesdropping (with cameras/microphone, wireless connections).

By design, Android™ is "open" to download apps from different sources. Its security depends on restricting apps by combining digital signatures, sandboxing, and permissions. Unfortunately, these restrictions can be bypassed, without the user noticing, by colluding apps for which combined permissions allow them to carry out attacks.

A basic example of collusion consists of one app permitted to access personal data, which passes the data to a second app allowed to transmit data over the network. While collusion is not a widespread threat today, it opens an avenue to circumvent Android™ permission restrictions that could be easily exploited by criminals to become a serious threat in the near future.

Almost all current research efforts are focusing on detection of single malicious apps. The threat of colluding apps is challenging to detect because of the myriad and possibly stealthy ways in which apps might communicate with each other. Existing Anti-Virus (AV) products are not designed to detect collusion. A review of the literature shows that detecting application collusion introduces a new set of challenges including: the detection of communication channels between apps, the exponential number of app combinations, and the difficulty of actually proving that two or more apps are really colluding.

1.2 Contribution

In this chapter we report on recent and ongoing research results from our ACID project[1] which suggest a number of reliable means to detect collusion, tackling the aforementioned problems. To this end we present our conceptual work on the topic of collusion and discuss a number of automated tools arising from it.

We start with an overview on the Android[TM] Operating System, which introduces the various security mechanism built in.

Then we give a definition for app collusion, and distinguish collusion from the closely related phenomena of collaboration and confused deputy attacks.

Based on this we address the exponential complexity of the problem by introducing a filtering phase. We develop two methods based on a lightweight analysis to detect if a set of apps has any collusion potential. These methods extract features through static analysis and use first order logic and machine learning to assess whether an analysed app set has collusion potential. By developing two methods to detect collusion potential we address the problem of collusion with two distinct approaches.

The first order logic approach allows us to define collusion potential through experts, which may identify attack vectors that are not yet being seen in the real world. Whereas the machine learning approach uses Android[TM] permissions to systematically assign the degree of collusion potential a set of apps may pose. A mix of techniques as such provides for an insightful understanding of possibly colluding behaviours and also adds confidence into filtering.

Once we have reduced the search space, we use a more computational intensive approach, namely software model checking, to validate the actual existence of collusion between the analysed apps. To this end, model checking provides dynamic information on possible app executions that lead to collusion; counter examples (witness traces) are generated in such cases.

---

[1] http://acidproject.org.uk.

In order to evaluate our approaches, we have developed a set of specifically crafted apps and gathered a data set of more than 50,000 real-world apps. Some of our approaches have been validated by using the crafted app set and tested against the real-world apps.

The above effort has been demonstrated through a number of publications through the different teams involved focusing on different aspects. The chapter allows us to provide a consolidated perspective on the problem. By systematically taking related work into account, we aim to provide a comprehensive presentation of state of the art in collusion analysis.

This chapter reports previously published work [6–8, 11, 13, 37], however, we expand on these publications by providing more detail and also by putting the singular approaches into context.

2 The Android™ Operating System

The Android™ operating system consists of three software layers above a Linux® kernel as shown in Fig. 1. The Linux® kernel is slightly modified for an embedded environment. It runs device-specific hardware drivers, and manages power and the file system. Android™ is agnostic of the processor (ARM, x86, and MIPS) but does take advantage of some hardware-specific security capabilities, e.g., the ARM v6 eXecute-Never feature for designating a non-executable memory area.

Above the kernel, libraries of native machine code provide common services to apps. Examples include Surface Manager for graphics; WebKit® for browser rendering; and SQLite for basic datastore. In the same layer, each app runs in its own instance of Android™ runtime (ART) except for some apps that are native, e.g., core Android™ services. A replacement for the Dalvik virtual machine (VM) since Android™ 4.4, the ART is designed to run Dalvik executable (DEX) byte-code on resource-constrained mobile devices. It introduces ahead-of-time (AOT) compilation converting bytecode to native code at installation time (in contrast to the Dalvik VM which interpreted code at runtime).

Fig. 1 Android™ operating system layers

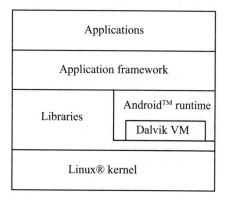

The application framework above the libraries offers packages and classes to provide common services to apps, for example: the Activity Manager for starting activities; the Package Manager for installing apps and maintaining information about them; and the Notification Manager to give notifications about certain events to interested apps.

The highest application layer consists of user-installed or pre-installed apps. Java source code is compiled into JAR (Java archive) files composed of multiple Java class files, associated metadata and resources, and optional manifest. JAR files can be translated into DEX bytecode and zipped into Android^{TM} package (APK) files for distribution and installation. APK files contain .dex files, resources, assets, lib folder of processor-specific code, the META-INF folder containing the manifest MANIFEST.MF and other files, and an additional AndroidManifest.xml file. The AndroidManifest.xml file contains the necessary configuration information to install the app, notably defining permissions to request from the user.

2.1 App Components

Android^{TM} apps are built composed of one or more components which must be declared in the manifest.

- *Activities* represent screens of the user interface and allow the user to interact with the app. Activities run only in the foreground. Apps are generally composed of a set of activities, such as a "main" activity launched when a user starts an app.
- *Services* operate in the background to carry out long-running tasks for other apps, such as listening to incoming connections or downloading a file.
- *Broadcast receivers* respond to messages that are sent through Intent objects, by the same or other apps.
- *Content providers* manage data shared across apps. Apps with content providers enable other apps to read and write their local data.

Any component can be public or private. If a component is public, components of other apps can interact with it, e.g., start the Activity, start the Service. If a component is private, only components from the app that runs with the same user ID (UID) can interact with that component.

2.2 Communications

Android^{TM} allows any app to start another app's component in order to avoid duplicate coding for the same function. However, this can not be done directly because apps are separate processes. To activate a component in another app, an app must deliver a message to the system that specifies the intent to start that component.

Intents are message objects that contain information about the operation to be performed and relevant data. Intents are delivered by various methods to application components, depending on the type of component. Intents about certain events are broadcasted, e.g., an incoming phone call. Intents can be explicit for specific recipients or implicit, i.e., broadcast through the system to any components listening. Components can provide Intent filters to specify which Intents a component is willing to handle.

Besides Intents, processes can communicate by standard Unix® communication methods (files, sockets), Android™ offers three inter-process communication (IPC) mechanisms:

- *Binder*: a remote procedure call mechanism implemented as a custom Linux® driver;
- *Services*: interfaces directly accessible using Binder;
- *Content Provider*: provide access to data on the device.

2.3 App Distribution and Installation

Android™ apps can be downloaded from the official Google Play™ market or many third party app stores. To catch malicious apps from being distributed, Google@ uses a variety of services including Bouncer, Verify Apps, and Safety Net. Since 2012, the Bouncer service automatically scans the Google Play™ market for potentially malicious apps (known malware) and apps with suspicious behaviours. It does not examine apps installed on devices or apps in third party app stores. Currently, however, none of these services look for apps exhibiting collusion behaviours.

The Verify Apps service scans apps upon installation on an Android™ device and scans the device in the background periodically or when triggered by potentially harmful behaviours, e.g., root access. It warns users of potentially harmful apps (PHAs) which may be submitted online for analysis.

The Safety Net service looks for network-based security threats, e.g., SMS abuse, by analyzing hundreds of millions of network connections daily. Google@ has the option to remotely remove malicious apps.

2.4 Android™ Security Approach

Android™ security aims to protect user data and system resources (including the network), which are regarded as the valuable assets. Apps are assumed to be untrusted by default and therefore considered potential threats to the system and other apps. The primary method of protection is isolation of apps from other apps, users from other users, and apps from certain resources. IPC is possible but mediated by Binder®. However, the user is ultimately in control of security by choosing which permissions to grant to apps. For more detailed information about Android™ security, the reader is referred to the extensive literature [21, 22, 27, 32, 59].

Android™ security is built on key security features provided by the Linux® kernel, namely: a user-based permissions system; process isolation; and secure IPC. In Linux®, each Linux® user is assigned a user (UID) and group ID (GID). Access to each resource is controlled by three sets of permissions: owner (UID), group (GID), and world. The kernel isolates processes such that users can not read another user's files or exhaust another's memory or CPU resources. Android™ builds on these mechanisms. An Android™ app runs under a unique UID, and all resources for that app are assigned full permissions for that UID and no permissions otherwise. Apps can not access data or memory of other apps by default. A user with root UID can bypass any permissions on any file, but only the kernel and a small subset of core apps run with root permission.

By default, apps are treated as untrusted and can not interact with each other and have limited access to the operating system. All code above the Linux® kernel (including libraries, application framework, and app runtime) is run within a sandbox to prevent harm to other apps or the system.

Apps must be digitally signed by their creators although their certificate can be self signed. A digital signature does not imply that an app is safe, only that the app has not been changed since creation and the app creator can be identified and held accountable for the behaviour of their app. A permissions system controls how apps can access personal information, sensitive input devices, and device metadata. By default, apps collecting personal information restricts data access to themselves. Access to sensitive user data is available only through protected APIs. Other types of protected APIs include: cost sensitive APIs that might generate a cost to the user; APIs to access sensitive data input devices such as camera and microphone; and APIs to access device metadata. App permissions are extracted from the manifest at install time by the PackageManager.

The default set of Android™ permissions is grouped into four categories as shown in Table 1.

The permissions system has known deficiencies. First, apps tend to request for excessive permissions. Second, users tend to grant permissions to apps without fully understanding the permissions or their implications in terms of risk. Third, the permissions system is concerned only with limiting the actions of individual apps. It is possible for two or more apps to collude for a malicious action by combining their permissions, even though each of the colluding apps is properly restricted by permissions.

3   App Collusion

ISO 27005 defines a threat as "A potential cause of an incident, that may result in harm of systems and organisations." For mobile devices, the range of such threats includes [62]:

Table 1 The default set of Android™ permissions

Category	Description	Examples
Normal	Can not cause real harm	ACCESS_NETWORK_STATE INTERNET SET_WALLPAPER
Dangerous	Possibly causing harm	READ_CONTACTS ACCESS_FINE_LOCATION READ_PHONE_STATE
Signature	Automatically granted if the app is signed by the same digital certificate as the app that created the permission	ACCESS_VR_MANAGER WRITE_BLOCKED_NUMBERS BIND_TRUST_AGENT
SignatureOrSystem	Similar to Signature except automatically granted to the Android™ system image in addition to the requesting app	GET_APP_OPS_STATS MANAGE_DEVICE_ADMINS ACCESS_CACHE_FILESYSTEM

- Information theft happens when information is sent outside the device boundaries.
- Money theft happens, e.g., when an app makes money through sensitive API calls (e.g. SMS).
- Service or resource misuse occurs, for example, when a device is remotely controlled or some device function is affected.

As we have seen before, the Android™ OS runs apps in sandboxes, trying to keep them separate from each other, especially that no information can be exchanged between them. However, at the same time Android™ has communication channels between apps. These can be documented ones (overt channels), or undocumented ones (covert channels). An example of an overt channel would be a shared file or intent; an example of a covert channel would be volume manipulation (the volume is readable by all apps) in order to pass a message in a special code.

Broadly speaking, app collusion is when, in performing a threat, several apps are working together, i.e., they exchange information which they could not obtain on their own.

This informal definition is close to app collaboration, where several apps share information (which they could not obtain on their own), in order to achieve a documented objective.

A typical example of collusion is shown in Fig. 2, where two apps perform the threat of information theft: the Contact_app reads the contacts database to pass the data to the Weather_app, which sends the data outside the device boundaries. The information between apps is exchanged through shared preferences.

In contrast, a typical example of collaboration would be the cooperation between a picture app and an email app. Here, the user can choose a picture to be sent via email. This requires the picture to be communicated over an overt channel from the picture app to the email app. Here, the communication is performed via a shared image file, to which both apps have access.

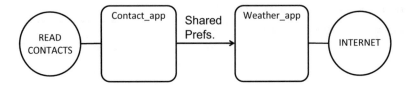

Fig. 2 An example of colluding apps

These examples show that the distinction between collusion and collaboration actually lies in the notion of intention. In the case of the weather app, the intent is malicious and undocumented, in the case of sending the email, the intent is documented, visible to the user and useful.

To sharpen the argument, it might be the case that the picture app actually makes the pictures readable by all apps, so that harm can be caused by some malicious app sending pictures without authorisation. This would provide a situation, where a bug or a vulnerability of one app is abused by another app, leading to a border case for collusion. In this case one would speak about "confused deputy" attack: the picture app has a vulnerability, which is maliciously abused by the other app, however, the picture app was—in the way we describe it here—not designed with the intention to collude. An early reference on such attacks is the work by Hardy [34].

This discussion demonstrates that notions such as "malicious", intent, and visibility (including app documentation—external and built-into the app) play a role when one wants to distinguish between collusion, cooperation, and confused deputy. This is typical in cyber security, see e.g. Harley's book chapter "Antimalware evaluation and Testing", especially the section headed "Is It or Isn't It?", [35, pp. 470–474]. It is often a challenge, especially for borderline cases, to distinguish between benign and malicious application behaviours. One approach is to use a pre-labeled "malicious" data set of APKs where all the aforementioned factors have been already accounted for. Many security companies routinely classify Android™ apps into clean and malicious categories to provide anti-malware detection in their products and we had access to such set from Intel Security (McAfee®). All apps classified as malicious fall into three mentioned threat categories. Now, collusion can be regarded as a camouflage mechanism applied to conceal these basic threat's behaviours. After splitting malicious actions into multiple individual apps they would easily appear harmless when checked individually. Indeed, even permissions of each such app would indicate it cannot pose a threat in isolation. But in combination, however, they may realise a threat. Taking into account all the details contributing to "maliciousness"—deceitful distribution, lack of documentation, hidden functionality, etc.—is practically impossible to formalise.

Here, in our book chapter, we aim to apply purely technical methods to discover collusion. Thus, we will leave out of our definition all aspects relating to psychology, sociology, or documentation. In the light of the above discussion our technical definition of collusion thus applies to all three identified cases, namely collusion, cooperation, and confused deputy. If someone aims to distinguish between these, then further manual analysis would be required, involving the distribution methods, documentation, and all other surrounding facts and details.

When analysing the APKs of various apps for collusion, we look at the actions that are being executed by these APKs. Actions are operations provided by the Android™ API (such as record audio, access file, write file, send data, etc.). We denote the set of all actions by *Act*. Note that this set also includes actions describing communication. Using an overt channel in Android™ requires an API call.

An action can be can be characterised by a number of static attributes such as permissions, e.g., when an app needs to record audio, the permission RECORD_AUDIO needs to be set in the manifest while the permission WRITE_EXTERNAL_STORAGE needs to be set for writing a file.

Technically, we consider a threat to be a sequence of actions. We consider a threat to be realised by collusion if it is distributed over several apps, i.e.,

Definition 1 there is a non-singleton set *S* of apps such that:

- each app in *S* contributes the execution of at least one action to the threat,
- each app in *S* communicates with at least one other app.

This definition will be narrowed down further when discussing concrete techniques for discovering collusion.

To illustrate our definition we present an abstract example.[2]

Example 1 (Stealing Contact Data) The two apps graphically represented in Fig. 2 perform information theft: the Contact_app reads the contacts database to pass the data to the Weather_app, which sends the data outside the device boundaries. The information is sent through shared preferences.

Using the collusion definition we can describe the actions performed by both apps as:

- $Act_{\text{Contact\_app}} = \{a_{read\_contacts}, send_{shared\_prefs}, \}$ and
- $Act_{\text{Weather\_app}} = \{a_{sendfile}, recv_{sharedprefs}, \}$

with the permissions $pms\,(a_{read\_contacts}) = \{Permission\_contacts\}$ and $pms(a_{send\_file}) = \{Permission\_internet\}$. The information threat T is given by

$$T = \langle a_{read\_contacts}, send_{shared\_prefs}, recv_{shared\_prefs}, a_{send\_file} \rangle.$$

This data leakage example is in line with the collusion definitions given in most existing work [5, 17, 40, 42, 46, 52] which regards collusion as the combination of inter-app communication with information leakage. However, our definition of a threat is broader, as it includes also financial and resource/service abuse.

---

[2]Concrete examples are available on request.

3.1   App Collusion in the Wild

We present our analysis of a set of apps in the wild that use collusion to maximise the effects of their malicious payloads [11]. To the best of our knowledge, this is the first time that a large set of colluding apps have been identified in the wild. This does not necessarily mean that there are no more colluding apps in the wild, as one of the main problems (that we are addressing in our work) is the lack of tools to identify colluding apps. We identified these sets of apps while looking for collusion potential on a set of more than 40,000 apps downloaded from App markets. While performing this analysis we found a group of apps that was communicating using both intents and shared preference files. A manual review of the flagged apps revealed that they were sharing information through shared preferences files to synchronise the execution of a potentially harmful payload. Both the colluding and malicious payload were included inside a library, the MoPlus SDK, embedded in all apps. This library has been known to be malicious since November 2015 [58]. However, the collusion behaviour of the SDK was hitherto unknown. In the rest of this section, we briefly describe this colluding behaviour.

The detected colluding behaviour looked different from the behaviour predicted by most app collusion research [47, 57] so far. In a nutshell, all apps including the MoPlus SDK that are running on a device will talk to each other to check which of the apps has the most privileges. This app will then be chosen to execute the local HTTP server able to receive commands from the C&C server, maximising the effects of the malicious payload.

The MoPlus SDK includes the MoPlusService and the MoPlusReceiver components. In all analysed apps, the service is exported. In Android™, this is considered to be a dangerous practice, as also other apps will be able to call and access this service. However, in this case it is a feature used by the SDK to enable communication between its apps.

The colluding behaviour is executed when the MoPlusService is created (onCreate method). This behaviour is triggered by the MoPlus SDK of each app and can be divided in two phases: establishing app priority and executing the malicious payload. To establish the app priority—see Fig. 3—the MoPlus SDK executes a number of checks, including the verifying if the app embedding the SDK has granted the INTERNET, READ_PHONE_STATE, ACCESS_NETWORK_STATE, WRITE_CONTACTS,   WRITE_EXTERNAL_STORAGE   or   GET_TASKS permissions.

After the priority has been obtained and stored, each service inspects the contents of the shared preference files to get its priority, returning the package name of the one with highest priority. Then, each service cancels previous intents being registered (to avoid launching the service more than once) and sends an intent targeting only the process with the higher previously saved priority—see Fig. 4.

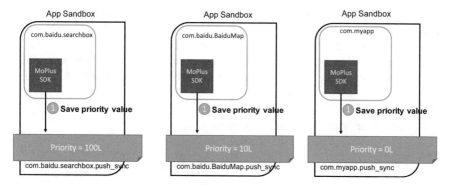

Fig. 3 Phase 1 of the colluding behaviour execution. Each app saves a priority value that depends on the amount of access it has to the system resources. Priority values are shown for the sake of explanation

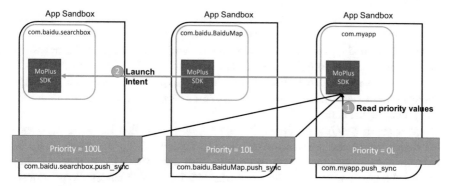

Fig. 4 Phase 2 of the colluding behaviour execution. Each app checks the WORLD_READABLE SharedPreference files and sends an intent to the app with highest priority

3.1.1 Discussion

It is important to notice that although all applications already include a malicious payload that could be executed on their own, if two apps with the Moplus SDK were to be installed in the same device, they would not be able to execute their individual malicious payloads. Although this assumption may seem unrealistic at first, implementing these kinds of behaviours inside SDKs makes this much more likely to happen. If we consider this assumption, then, the colluding behaviour allows two things: first, it enables the execution of the malicious payload avoiding concurrency problems between all instances of the SDK running. Second, it allows the SDK instance with highest access to resources to be the one executing, maximising the result of the attack. This introduces an important remark in how colluding applications have to be analysed. This is, having the static features that allow them to execute a threat doesn't mean they will be able to achieve that threat in all scenarios, like the one presented in our case. This means, that when considering app collusion we must look not only to the specific features or capabilities of the app, but also how those capabilities work when the app is being executed with other apps. If we are considering collusion it does not make much sense to consider the capabilities of an app in isolation with respect to other apps, we have to consider the app executing in an environment where there are other apps installed.

3.1.2 Relation with Collusion Definition

This set of apps found in the wild relates to our collusion definition in the following way. Consider a set of apps $S = \{app_1, app_2, \cdots, app_n\}$ that implements the MoPlus SDK. As they embed the MoPlus SDK, the attacks that can be achieved by them includes writing into the contacts database, launching intents and installing applications without user interaction among others. This set of threats was identified by TrendMicro researchers [58].

Consider now the installation of an application without the user interaction as a threat $T_{install}$. As all apps embed the MoPlus SDK, all apps include the code to potentially execute such threat, but only apps that request the necessary permissions are able to execute it. If app_i is the only app installed in the device, and has the necessary permissions, executing $T_{install}$ will require the following actions $\{Open\ server_i, Receive\ command_i, Install\ app_i\}$, the underscore being the app executing the action.

However, if another MoPlus SDK app, app_j, is installed in the same device but doesn't have the permissions required to achieve $T_{install}$ the threat won't be realised because of concurrency problems, both apps share the port where they receive the commands. To avoid these, the MoPlus SDK includes the previously described leader selection mechanisms that uses the SharedPreferences. In this setting, we can describe the set of actions required by both apps to execute the threat as $Act_{Moplus} = \{Check\ permissions_i, Check\ permissions_j, Save\ priority\ i_i,$ $Save\ priority\ j_j, Read\ priority\ i_j, Read\ priority\ j_i, Launch\ service\ i_j, Open\ server_i,$ $Receive\ command_i, Install\ app_i\}$. Considering $Read\ priority\ x_y$ and $Save\ priority\ x_y$ as actions that make use of the SharedPreferences as a communication channel, we can consider that the presented set of actions follows under our collusion definition as (1) there is a sequence of actions that execute a threat executed collectively by app_i and app_j and (2) both apps communicate with each other.

4 Filtering

A frontal attack on detecting collusions to analyse pairs, triplets and even larger sets is not practical given the search space. An effective collusion-discovery tool must include an effective set of methods to isolate potential sets which require further examination.

4.1 Rule Based Collusion Detection

Here, in a first step we extract information about app communications and access to protected-resources. Using rules in first order logic codified in Prolog, the method identifies sets of apps that might be colluding.

The goal of this is to serve as a fast, computationally cheap filter that detects potentially colluding apps. For such a first filter it is enough to be based on permissions. In practical work on real world apps this filter turns out to be effective to detect colluding apps in the wild.

Our filter (1) uses Androguard [20] to extract facts about the communication channels and permissions of all single apps in a given app set S, (2) which is then abstracted into an over-approximation of actions and communication channels that could be used by a single app. (3) Finally the collusion rules are fired if the proper combinations of actions and communications are found in S.

4.1.1 Actions

We utilise an action set Act_{prolog} composed out of four different high level actions: accessing sensitive information, using an API that can directly cost money, controlling device services (e.g. camera, etc.), and sending information to other devices and the Internet. To find out which of these actions an app could carry out, we extract its set of permissions pms_{prolog} with Androguard. For each found permission, our tool creates a new Prolog fact in the form $uses(app, permission)$. Then permissions extracted are mapped to one of the four high level actions. This is done with a set of previously defined Prolog rules. The mapping of all Android™ permissions to the four high-level actions can be found in the project's Github repository.[3] As an example, an app that declares the INTERNET permission will be capable of sending information outside the device:

$$uses(App, P_{Internet}) \rightarrow information\_outside(App)$$

4.1.2 Communications

The communication channels established by an app are characterised by its API calls and the permissions declared in its manifest file. We cover communication actions (com_{prolog}) that can be created as follows:

- *Intents* are messages used to request tasks from other application components (activities, services or broadcast receivers). Activities, services and broadcast receivers declare the intents they can handle by declaring a set of intent filters.

---

[3]https://github.com/acidrepo/collusion_potential_detector.

- *External Storage* is a storage space shared between all the apps installed without restrictions. Apps accessing the external storage need to declare the

READ_EXTERNAL_STORAGE

  permission. To enable writing, apps must declare

WRITE_EXTERNAL_STORAGE.

- *Shared Preferences* are an OS feature to store key-value pairs of data. Although it is not intended for inter-app communication, apps can use key-value pairs to exchange information if proper permissions are defined (before Android^{TM} 4.4).

We map apps to sending and receiving actions by inspecting their code and manifest files. When using intents and shared preferences we are able to specify the communication channel using the intent actions and preference files and packages respectively. If an application sends a broadcast intent with the action SEND_FILE we consider the following:

$$send\_broadcast(App, Intent_{send\_file})$$
$$\rightarrow send(App, Intent_{send\_file})$$

We consider that two apps communicate if one of them is able to *send* and the other to *receive* through the same channel. This allows to detect communication paths composed by an arbitrary number of apps:

$$send(App_a, channel) \wedge receive(App_b, channel) \rightarrow$$
$$communicate(App_a, App_b, channel)$$

4.1.3   Collusion Potential

To identify collusion potential in app sets, we put together the different communication channels found in an app and their high-level actions as identified by their permissions. Then, using domain knowledge we created a threat set that describes some of the possible threats that could be achieving with a collusion attack. Our threat set τ_{prolog} considers information theft, money theft and service misuse. As our definition states, each of the threats is characterised by a sequence of actions. In fact, each of our collusion rules gathers the two elements required by the collusion definition explained in Sect. 3: (1) each app of the group must execute at least one action of the threat and (2) each app in S communicates at least with another app in S. The following rule provides the example of an information threat executed through two colluding apps:

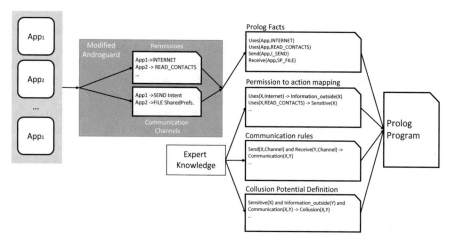

Fig. 5 General overview of the process followed in the rule based collusion detection approach

$$sensitive\_information(App_a)$$
$$\wedge \; information\_outside(App_b)$$
$$\wedge \; communicate(App_a, App_b, channel)$$
$$\rightarrow \quad collusion(App_a, App_b)$$

Note that more apps could be involved in this same threat as simply forwarders of the information extracted by the first app until it arrives to the exfiltration app. This case is also covered by Definition 1, as the forwarding apps need to execute their communication operations to succeed on their attack (fulfilling both of our definition conditions).

Finally, the Prolog rules defining collusion potential, the facts extracted from apps, and rules mapping permissions to high level actions and communications between apps are then put on a single file. This file is then fed into Prolog so collusion queries can be made. The overall process is depicted in Fig. 5.

4.2 Machine Learning Collusion Detection

Security solutions using machine learning employ algorithms designed to distinguish between malicious and benign apps. To this end, they analyse various features such as the APIs invoked, system calls, data-flows and resources used assuming a single malicious app attack model. In this section, we extend the same notion to assess the collusion threat which serves as an effective filtering mechanism for finding collusion candidates of interest. We employ probabilistic generative modelling for this task with the popular Naive Bayesian model.

4.2.1 Naive Bayes Classifier

Let $X = [x_1, \ldots, x_k]$ be a k-dimensional space with binary inputs, where k is the total number of permissions in Android™ OS and $x_j \in \{0, 1\}$ are independent Bernoulli random variables. A variable x_j takes the value 1 if permission j is found in the set S of apps under consideration, 0 otherwise. Let $Y = \{m\text{-malicious}, b\text{-benign}\}$ be a one dimensional output space. The generative naive Bayes model specifies a joint probability, $p(x, y) = p(x \mid y).p(y)$, of the inputs x and the label y: the probability $p(x, y)$ of observing x and y at the same time is the same as the probability $p(x \mid y)$ of x happening when we know that y happens multiplied with the probability $p(y)$ that y happens. This explicitly models the actual distribution of each class (i.e. malicious and benign in our case) assuming that some parameters stochastic process generates the observations, and hence estimates the model parameters that best explains the training data. Once the model parameters are estimated (say $\hat{\theta}$), then we can compute $p(t_i \mid \hat{\theta})$ which gives the probability of the ith test case is generated by the derived model. This can be applied in a classification problem as explained below.

Let p(x,y) be a joint distribution over $X \times Y$ from which a training set $R = \{x_i^k, y_i^1 \mid i = 1, 2, 3, \ldots, n\}$ of n independent and identically distributed examples are drawn. The generative naive Bayes classifier uses R to estimate $p(x|y)$ and $p(y)$ in the joint distribution. If $c(.)$ stands for counting the number of occurrences of an event in the training set,

$$\hat{p}(x = 0 \mid y = m) = \frac{c(x = 0, y = m) + \alpha}{c(y = m) + 2\alpha} \tag{1}$$

where the pseudo count $\alpha > 0$ is the smoothing parameter. If $\alpha = 0$, i.e. taking the empirical estimates of the probabilities without smoothing, then

$$\hat{p}(x = 0 \mid y = m) = \frac{c(x = 0, y = m)}{c(y = m)} \tag{2}$$

Equation (2) estimates the likelihoods using the training set R. Uninformative priors, i.e. $\hat{p}(y = m)$, can also be estimated in the same way. Instead, we estimate prior distribution in an informative way in this work as it would help us in modelling the knowledge not available in data (e.g. permission's critical level). Informative prior estimation is described in Sect. 4.2.3.

In order to classify the ith test case, the model predicts $p(t_i \mid \hat{\theta}) = m$ if and only if:

$$\frac{\hat{p}(x = t_i, y = m)}{\hat{p}(x = t_i, y = b)} > 1 \tag{3}$$

4.2.2 Threat Likelihood

As per our collusion definition in Sect. 3, estimating the collusion threat likelihood $L_c(S)$ of a non-singleton set S of apps involves two likelihood components $L_\tau(S)$ and $L_{com}(S)$: $L_\tau(S)$ expresses how likely the app set S can fulfil the sequence of actions required to execute a threat; $L_{com}(S)$ is the ability to communicate between apps in S. Using the multiplication rule of well-known basic principles of counting:

$$L_c(S) = L_\tau(S) \times L_{com}(S) \tag{4}$$

As mentioned above, we employ the so-called Naive Bayesian informative [50] model to demonstrate the evaluation of Eq. (4). First, we define the model, then train and validate the model, and finally test it using a testing data set.

4.2.3 Estimating L_τ

Let $X = [x_1, \ldots, x_k]$ be a k-dimensional random variable as defined in Sect. 4.2.1. Then the probability mass function $P(X)$ gives the probability of obtaining S with permissions as described in X. Our probabilistic model $P(X)$ is then given by Eq. (5):

$$P(X) = \prod_{j=1}^{k} \lambda_j^{x_j} (1 - \lambda_j)^{1-x_j} \tag{5}$$

where $\lambda_j \in [0, 1]$ is a Bernoulli parameter. In order to compute L_τ for a given set S, we define a sample statistic as $Q_s = \frac{\ln\{(P(X))^{-1}\}}{|S|}$, divide $\ln\{(P(X))^{-1}\}$ by the number of distinct permissions in set S, and scale down it to the range [0,1] by dividing the $max(Q_s)$ which estimated using empirical data. Hence, for a given set S, $L_\tau = \frac{Q_s}{max(Q_s)}$. The desired goal behind the above mathematical formulation is to make requesting more critical permissions to increase the likelihood of "being malicious" than requesting less critical ones regardless of frequencies. Readers who require a detailed explanation of the mathematical notion behind the above formulation are invited to refer to [50].

To complete our modelling, we need to estimate values $\hat{\lambda}_j$ that replace λ_j in the computation of L_τ. To this end—to avoid over fitting $P(X)$—we estimate λ_j using informative beta prior distributions [41] and define the maximum posterior estimation

$$\hat{\lambda}_j = \frac{\sum x_j + \alpha_j}{N + \alpha_j + \beta_j} \tag{6}$$

where N is the number of apps in the training data set and α_j, β_j are the penalty effects. In this work we set $\alpha_j = 1$. The values for β_j depend on the critical level of permissions as given in [50, 55]. β_j can take either the value $2N$ (if permission j is most critical), N (if permission j is critical) or 1 (if permission j is non-critical).

4.2.4   Estimating L_{com}

In order to materialise a collusion, there should be an inter app communication
closely related to the target threat. To establish this association we need to consider
a number of factors including the contextual parameters. At this stage of the
research we do not focus on estimating the strength of connection (association)
between the threat and the communication. Instead we investigate what percentage
of communication channels can be detected through static code analysis, and simply
assume[4] these channels can be used for malicious purpose by apps in set S. Hence
we consider L_{com} to be a binary function such that $L_{com} \in \{1, 0\}$ which takes the
value 1 if there is inter app communication within S using either intents or external
storage (we do not investigate other channels in this work).

4.2.5   Proposed Probabilistic Filter

Our probabilistic filter consists of two sub filters: an inner and an outer one. The
inner filter applies on the top of the outer filter. The outer filter is based on the L_τ
value which we can compute using permissions only. Permissions are very easy and
cheap to extract from APKs—no decompilation, reverse engineering, complex code
or data flow analysis is required. Hence the outer filter is computationally efficient.
The majority of non-colluding app pairs in an average app set can be treated using
this filter only (see Fig. 6). This avoids the expensive static/dynamics analysis on
these pairs. The inner filter is based on L_{com} value which we currently compute
using static code analysis. A third party research prototype tool Didfail [15] was
employed in finding intent based inter app communications. A set of permission
based rules was defined to find communication using external storage. Algorithm 1
presents the proposed filter to find out colluding candidates of interest.

Fig. 6 Validation: L_τ score
obtained by each pair in the
validation data set

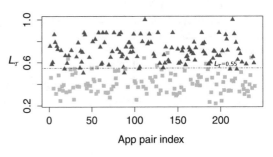

▲ Colluding pairs   ▪ Non-colluding pairs

---

[4]This assumption might produce false positives, however, never false negatives. It is left as a future
work to improve this.

Algorithm 1: Probabilistic filter. The outer filter is based on L_τ and the inner filter is based on L_{com}

Λ: Set of individual apps;
Ω: Set of pairs of colluding candidates of interest;
input : $\Lambda = \{app_1, app_2, app_3, \ldots, app_n\}$
output: $\Omega = \{pair_1, pair_2, pair_3, \ldots, pair_m\}$
if $|\Lambda| \geq 2$ **then**
    Let Θ = set of all possible app pairs in Λ;
    foreach $pair_j$ in Θ **do**
        Compute L_τ as described in section 4.2.3;
        /* outer filter   */
        if $L_\tau \geq$ threshold **then**
            Compute L_{com} as described in section 4.2.4 ;
            /* inner filter   */
            if $L_{com} == 1$ **then**
                Return $(pair_j)$;
            end
        end
    end
end

4.2.6 Experimental Setup and Validation

Algorithm 1 was automated using R^5 and Bash scripts. As mentioned above, it also includes calls to a third party research prototype [15] to find intent based communications in computing L_{com}. The model parameters in Eq. (5) were estimated using training datasets produced from a 29k size app sample provided by Intel security.

Our validation data set consists of 240 app pairs in which half (120) of them are known colluding pairs while the other half are non-colluding pairs. In order to prevent over fitting, app pairs in the validation and testing sets were not included in the training set. As shown in Fig. 6, the proposed method assigns higher L_τ scores[6] for colluding pairs than clean pairs. Table 2 presents the confusion matrix obtained for the proposed method by fitting a linear discriminant line (LDL), i.e. the blue dotted line in Fig. 6 (Sensitivity $= 0.95$, specificity $= 0.94$, precision $= 0.94$ and the F-score $= 0.95$).

However as shown in Fig. 6, colluding and non-colluding are not easily separable two classes by a LDL. There are some overlaps between class elements. As a result it produces false classifications in Table 2. It is possible to reduce false alarms by changing the threshold. For example either setting the best possible discriminant line or its upper bound (or even higher, see Fig. 6) as the threshold will produce zero false

---

[5]http://www.r-project.org/.

[6]We plot L_τ values in Fig. 6 as outer filter in Algorithm 1 depends on it, and to show that majority of non-colluding app pairs can be treated using L_τ only. However, it should be noted that $L_c = L_\tau$ for colluding pairs as $L_{com} = 1$.

Table 2 Confusion matrix for the naive Bayesian method

n = 240	Actual colluding	Actual non-colluding
Predicted colluding	114	7
Predicted non-colluding	6	113

positives or vice versa in Table 2. But as a result it will increase false negative rate that will affect on the F-score—the performance measure of the classifier. Hence it would be a trade-off between a class accuracy and overall performance. However since the base rate of colluding apps in the wild is close to zero as far as anyone knows, the false positive rate of this method would have to be vanishingly small to be useful in practice. Instead of LDL, using a non-linear discriminator would also be another possibility to reduce false alarms. This is left as a future work to investigate.

The average processing time per app pair was 80s which consists of $\leq 1s$ for the outer filter and rest of the time for the inner filter. Average time was calculated on a mobile workstation with an Intel Core i7-4810MQ 2.8 GHz CPU and 32 GB of RAM.

4.3   Evaluation of Filtering

We validate both our filtering methods against a known ground truth by applying them to a set of artificially created apps. Furthermore, we report on managing complexity by scaling up our rule based detection method to deal with 50,000+ real world applications.

4.3.1   Testing the Prolog Filter

The validation of the Prolog filter has been carried out with fourteen artificially crafted apps that cover information theft, money theft and service misuse. Created apps use Intents, Shared Preferences and External storage as communication channels. They are organised in four colluding sets:

- The **Document Extractor** set consists of one app (*id 1*) that looks for sensitive documents on the external storage; the other app (*id 2*) sends the information received (via SharedPreferences) to a remote server.
- The **Botnet** set consists of four apps. One app (*id 3*) acts as a relay that receives orders from the command and control center. The other colluding apps execute commands (delivered via BroadcastIntents) depending on their permissions: sending SMS messages (*id 4*), stealing the user's contacts (*id 5*) and starting and stopping tasks (*id 6*).

Table 3 Collusion matrix of the Prolog program

id	1	2	3	4	5	6	7	8	9	10	11	12	13	14
1		♣								♣	♣			
2														
3				$	♠	♠								
4														
5		♣								♣	♣			
6		♣	♣											
7	♣							♣	♣	♣	♣			
8									♣					
9														
10											♣			
11														
12													♣	♣
13														
14														

♣ = Information theft. $ = Money theft. ♠ = Service misuse. ♣, ♠, ♣ = False positives

- The **Contact Extractor** set consists of three apps. The first (*id 7*) reads contacts from the address book, the second (*id 8*) forwards them via the external storage to the third one (*id 9*), which sends them to the Internet. The first and second app communicate via BroadcastIntents .
- The **Location Stealing** set consists of one app (*id 12*) that reads the user location and shares it with the second app (*id 13*), which sends the information to the Internet.

The three non-colluding apps are a document viewer (*id 10*), an information sharing app (*id 11*) and a location viewer (*id 14*). The first app is able to display different file types in the device screen and use other apps (via broadcast intents) to share their uniform resource identifier (URI). The second app receives text fragments from other apps and sends them to a remote server. The third app receives a location from another app (with the same intent used by apps 12 and 13) and shows it to the user on the screen.

Table 3 shows the results obtained with our rule based approached. The entry "dark red club" in row 1 and column 2 means: the program detects that app *id 1* sends information to app *id 2*, and these two apps collude on an "information theft". As we take communication direction into consideration, the resulting matrix is non-symmetric, e.g., there is no entry in row 2 and column 1. The entry "light red club" in row 1 and column 10 means: the program flags collusion of type "information theft" though the set {*id 1*, *id 10*} is clean.This provides further information about the collusion attack. For instance, one can see the information leak in information theft attacks. Additionally, the way we defined the communication rules makes it possible to identify transitive collusion attacks (i.e. app 7 colluding with app 9 through app 8). The approach identifies all colluding app sets. It also flags eight false positives due to over-approximation. Note, that there are no false negatives due to the nature of our test set: it utilises only those communication methods that our Prolog approach is able to identify.

Our false positives happen mainly because two reasons. First, we do not consider in our initial classification some of the communication channels that are already widely use by apps in Android™. For example, the Intent with action VIEW or SEND are very common in Android™ applications. It is unlikely that apps would use them for collusion as other apps could have registered to receive the same information. Second, in this approach, we identify apps that are communicating by sharing access to sensitive resources, but we do not look at how that access is shared. It must be noted, that the main aim of this approach is to reduce the amount of app combinations that are being passed through the data-flow analysis.

4.3.2 Testing the Probabilistic Filter

We tested the Probabilistic filter with a different sample consisting of 91 app pairs. Figure 7 presents the outcome for this set. Each cell in the table denotes a L_τ value for the corresponding app pair. Note that though there are 196 possible pairs (i.e. 14×14 cells in the table), for readability, we leave the lower half empty since the table is symmetric. Pairs on the diagonal are also not interesting to our discussion. To minimise false negatives, we use the lower bound ($=0.50$) gained from the validation dataset for the discriminant line as threshold for L_τ. We report possible collusion if $L_\tau \geq 0.5$ and $L_{com} = 1$, otherwise we report non-collusion. Dark

ID	1	2	3	4	5	6	7	8	9	10	11	12	13	14
1		0.51	0.61	0.97	1	0.8	1	0.81	0.77	0.77	0.77	0.44	0.44	0.95
2			0.48	0.62	0.55	0.49	0.55	0.58	0.51	0.51	0.58	0.31	0.31	0.49
3				0.69	0.64	0.56	0.64	0.48	0.61	0.61	0.72	0.41	0.41	0.58
4					1	0.84	1	0.85	0.71	0.71	0.82	0.56	0.56	0.95
5						0.84	1	0.86	0.67	0.67	0.82	0.47	0.47	1
6							0.84	0.68	0.58	0.58	0.65	0.43	0.43	0.78
7								0.86	0.67	0.67	0.82	0.47	0.47	1
8									0.51	0.51	0.58	0.31	0.31	0.77
9										0.77	0.77	0.44	0.44	0.61
10											0.77	0.44	0.44	0.61
11												0.47	0.47	0.73
12													0.47	0.41
13														0.41
14														

Fig. 7 Testing the proposed filter. For readability—we leave the upper half empty since the table is symmetric. Pairs on the diagonal are also not interesting to our discussion. *Dark red* shows true positives, *light red* shows false positives, *dark green* shows true negatives, and *light green* shows false negatives

red shows true positives, light red shows false positives, dark green shows true negatives, and light green shows false negatives.

With regards to false alarms, app pair (1,2) was not detected by our analysis due to the third party tool does not detect communication using SharedPreferences. Since we do only pairwise analysis, app pair (7,9) was not reported. That pair depends on transitive communication. Pair (12,13) was not reported since L_τ is less than the chosen threshold. As mentioned in Sect. 4.2.6, it would be possible to reduce false alarms by changing the LDL threshold, but subject to degrading the overall performance measure of the classifier.

Precise estimation of L_{com} would be useful to reduce false alarms in our analysis. But it should be noted that existence of a communication is just a necessary condition to happen a collusion, however not a sufficient condition to detect it. In this context it is worth to mention that a recent study [23] shows that 84.4% of non-colluding apps in the market place can communicate with other apps either using explicit (11.3%) or implicit (73.1%) intent calls. Therefore the threat element (i.e. L_τ) is far more informative in collusion estimation than the communication element (L_{com}) in our model.

Both validation and testing samples are blind samples and we have not properly investigated them for the biasedness or realisticity.

5   Model-Checking for Collusion

Filtering is an effective method to isolate app sets. Using software model checking, we provide a sound method for proving app sets to be clean that also returns example traces for potential collusion based on the $\mathbb{K}$ framework [54]—c.f. Fig. 8. We start with a set of apps in the form of an Application Package File (APK). The DEX code in each APK file is disassembled into the Smali format with open source tools. The Smali code of the apps is parsed by the $\mathbb{K}$ tool. Compilation in the $\mathbb{K}$ tool translates the $\mathbb{K}$ representation of the Android™ apps into a rewrite theory in Maude [18]. Finally, the Maude model checker searches the transition system compiled by the $\mathbb{K}$ tool to provide an answer if the input set of Android™ apps colludes or not. In the case when collusion is detected, the tool provides a readable counter example trace. In this section we focus on information theft only.

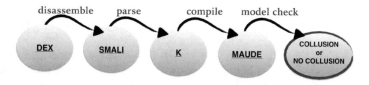

Fig. 8 Work-flow for the Android™ formal semantics in the $\mathbb{K}$ framework

5.1 Software Model Checking

Software model checking is a methodology widely used for the verification of properties about programs w.r.t. their executions. A profane view on model checking would be to see it as instance of the travelling salesman problem: every state of the system shall be visited and checked. This means that, upfront, model checking is nothing but a specialised search in a certain type of graph or, as it is known in the field, a *transition system*.

Initially, the application of model checking focused on simple transition systems, especially coming from hardware. *Simplicity* was necessary to contain a notorious efficiency problem known as the "state space explosion". Namely, the methodology would fail to produce timely efficient results due to the exponential nature of the complexity of the model checking procedures w.r.t. the number of system states.

Modern model checking tools attempt to meet the challenge posed by (higher level) programs, i.e., software, that are known to quickly produce a large (potentially unbounded) number of states, e.g., due to dynamic data structures, parallelism, etc. Hence, software model checking uses, in addition to basic model checking, other techniques (e.g. theorem proving or abstract interpretation) in order to coherently simplify the transition system given to the model checker.

A standard example is given by imperative programming languages. Here, a program p is viewed as a sequence of program locations $pc_i, i \geq 0$, that identify instructions. The effect of an instruction I at pc_i is a relation r_i which associates the states before with the states after the execution of I. Software model checking computes the transitive closure R of the relations r_i to obtain the set of *reachable* states of the program p.

Note, however, that for infinite state programs the computation of R may not terminate or may require an unreasonable amount of time or memory to terminate. Hence software model checking transforms the state space of the program into a "simpler" one by, essentially, eliminating unnecessary details in the relation r_i thus obtaining an *abstract* program a defined by the relations $a(r_i)$. The model checking of a, usually named "abstract" model checking, trades off precision for efficiency. A rigorous choice of the abstract set of states (i.e. abstract domain) and the abstract relations $a(r_i)$ (i.e. abstract semantics) ensures that the abstract model checking is sound (i.e. proving the property in the abstract system implies the property is proved in the original, concrete, system).

5.1.1 Challenges

In the following we will explain how we define a transition system using $\mathbb{K}$ and what abstractions we define in order to allow for an effective check for collusion.

Formalising Dalvik Byte-code in $\mathbb{K}$ poses a number of challenges: there are about 220 instructions to be formalised, the code is object oriented, it is register based (in contrast to stack based, as Java Byte-code), it utilises callbacks and intent based communication, see [3]. We provide two different semantics for DEX code, namely a concrete and an abstract one. While the concrete semantics has the benefit to be intuitive and thus easy to be validated, it is the abstract semantics that we employ for app model checking. We see the step from the descriptive level provided by [3] to the concrete, formal semantics as a 'controllable' one, where human intuition is able to bridge the gap. In future work, we intend to justify the step from the concrete semantics to the abstract one by a formal proof. Our implementation of both Android™ semantics in $\mathbb{K}$ is freely available.[7] The code of the colluding apps discussed in this section is accessible via an encrypted web-page. The password is available on request.[8]

5.2   The $\mathbb{K}$ Framework

The $\mathbb{K}$ framework [54] proposes a methodology for the design and analysis of programming languages; the framework comes with a rewriting-based specification language and tool support for parsing, interpreting, model-checking and deductive formal verification. The ideal work-flow in the $\mathbb{K}$ framework starts with a formal and executable language syntax and semantics, given as a $\mathbb{K}$ specification, which then is tested on program examples in order to gain confidence in the language definition. Here, the $\mathbb{K}$ framework offers model checking via compilation into Maude programs (i.e., using the existing reachability tool and LTL Maude model checker).

A $\mathbb{K}$ specification consists of *configurations*, *computations*, and *rules*, using a specialised notation to write semantic entities, i.e., $\mathbb{K}$-cells. For example, the $\mathbb{K}$-cell representing the set of program variables as a mapping from identifiers *Id* to values *Val* is given by $\langle Id \mapsto Val \rangle_{vars}$. Configurations in $\mathbb{K}$ are labelled and nested $\mathbb{K}$-cells, used to represent the structure of the program state. Rules in $\mathbb{K}$ are of two types: computational and structural. Computational rules represent transitions in a program execution and are specified as configuration updates. Structural rules provide internal changes of the program state such that the configuration can enable the application of computational rules.

5.3   A Concrete Semantics for Dalvik Code

The concrete semantics specifies system configurations and transition rules for all Smali instructions and a number of Android™ API calls in $\mathbb{K}$. Here, we strictly follow their explanation [2].

Fig. 9 Android™ configuration

5.3.1 System Configurations

Configurations are defined in $\mathbb{K}$ style as cells which might contains sub-cells. Top of a configuration is a "sandboxes" cell, containing a "broadcasts" sub-cell abstracting the Android™ intent broadcast service and possibly multiple "sandbox" cells capturing states of installed apps (Fig. 9).

In $\mathbb{K}$, the asterisk symbol next to the name "sandbox" specifies that the number of "sandbox" cells within a "sandboxes" cell is 0 or more. Each sandbox cell simulates the environment in which an application is isolated. It contains the classes of the application, the currently executed thread, and the memory storing the objects that have been instantiated so far. For the current thread, we store the instructions left to be run in a "k" cell, while the content of the current registers are kept in a "regs" cell . Classes and Method cells can be defined similarly. In turn, each "method" cell consists of the name of the method, the return type of the method and the statements of the method within a "methodbody" cell. Finally, "object" cells are used to store the objects that have been instantiated so far. They are stored within the "memory" cell of a "sandbox". As depicted in Fig. 10, an object cell contains a reference (an integer), its type, values of its corresponding fields, a Boolean value to indicate whether the object is sensitive and the set of applications that have created this object. The last two cells have been added for the sake of program analysis.

5.3.2 Smali Instructions

As a concrete example of how to formalise an instruction, let us consider the iget $R1, R2, CN \rightarrow FN : FT$ instruction. iget retrieves the field value of an object. Here, CN is the name of a class, FN and FT are the name of the field to be read and its type, register $R2$ holds the reference to the object to be read from, and—after execution of the instruction—register $R1$ shall hold the value of the field FN. The $\mathbb{K}$ rule for its semantics is illustrated in Fig. 11. This $\mathbb{K}$ rule is enabled when (1) the k cell of a thread starts with an iget instruction, (2) $R2$ is resolved to a reference $I2$ of some object where (3) FN maps to a value of $TV1$. When the rule is applied, $TV1$ is copied into $R1$.

The semantics for Smali instructions in $\mathbb{K}$ is organised in a number of separate modules as shown in Fig. 12, where arrows specify import. The "semantic-core" contains the semantics rules for basic instructions and directives such as "nop" (no operation), ".registers n" and ".locals n" where n is an integer. Additionally,

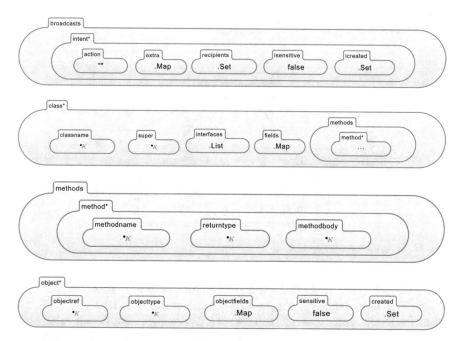

Fig. 10 Sub-cells of a configuration: broadcasts and object

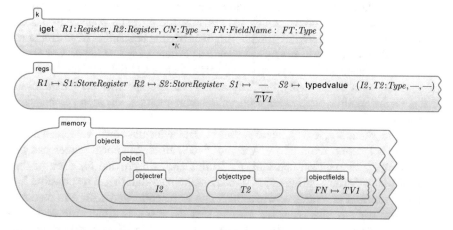

Fig. 11 $\mathbb{K}$ rule for the semantics of `iget` instruction

it also defines several auxiliary functions which are used later in other modules for semantic rules. For example, the function "isKImplementedAPI" is defined to determine whether an API method has been implemented within the $\mathbb{K}$ framework; if not, the interpreter will look for it within the classes of the invoking application.

The "loading" module is responsible for constructing the initial configuration. When running a Smali file in the $\mathbb{K}$ framework, it will parse the file according

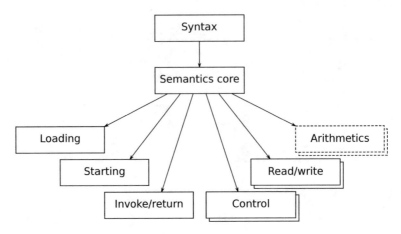

Fig. 12 Semantic module structure

to the defined syntax and place the entire resulting abstract syntax tree (AST) in a cell. The rules defined in the loading module are used to take this AST and distribute its elements to the right cells of the initial configuration. In particular, each application is placed in a sandbox cell, its classes are placed in the classes cell, etc. The "invoke/return" module defines the semantic rules for invoking methods and return instructions. The "control" module specifies the semantics of instructions such as "if-then" and "goto", which may change the program counter in a non-trivially way. The "read/write" module implements the semantics of instructions for manipulating objects in the memory such as instantiating new objects or array, initialising elements of an array, retrieving value of an object field and changing the value of an object field. Finally, the "arithmetics" module specifies the semantics of arithmetic instructions such as addition, subtraction, multiply, division and bit-wise operations.

In some situations, our semantics has to deal with unknown values such as the device's location returned by Android™ OS. In $\mathbb{K}$, unknown values can be represented by the built-in constant $\cdot K$. To this end, we provide for each of the "control", "read/write", "arithmetics" modules a counter-part that is responsible for unknown values. For example, when the value to be compared with 0 in an `ifz` Smali instruction is unknown, we assume that the result is either true or false, thereby leading to a non-deterministic execution of the Smali program. Similarly, arithmetical operations propagate unknown values.

5.3.3 Semantics for the Android™ APIs

Regarding the semantics of the Android™ APIs which encompasses a rich set of predefined classes and methods, API classes and methods usually come together with Android™ OS on an Android™ device and hence are not included in the DEX

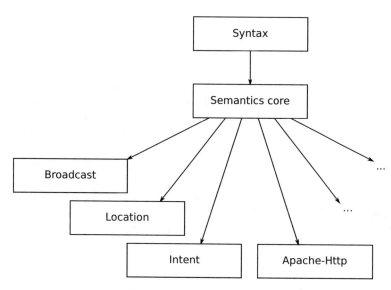

Fig. 13 Semantic module structure for Android™ API

code of an app. Obviously, one may obtain the Smali code of those API classes and methods. However, this will significantly increase the size of the Smali code to be analysed in $\mathbb{K}$ and consequently the state space of the obtained models. To this end, we directly implement the semantics of some of these classes and methods in $\mathbb{K}$ rules, based on their description [2]. While the first approach appears to be more faithful, it would significantly increase the size of the Smali code to be analysed in $\mathbb{K}$ and consequently the state space of the obtained models. This is avoided by the second approach where one can choose the abstraction level required for the analysis in question.

In Fig. 13, we show the structure for $\mathbb{K}$ modules which implements the semantics of some API methods.

In particular, we have implemented a number of APIs, including modules Location, Intent, Broadcast, and Apache-Http. Other API classes and methods can be implemented similarly. For those modules that are not (yet) implemented in $\mathbb{K}$, we provide a mechanism that a call to any of them returns an unknown result, i.e., the "$\cdot K$" value.

A typical example is the Location module which is responsible for implementing the semantics of API methods relating to the Location Manager such as registering a callback function when the device's location changes, i.e., the `requestLocationUpdates` method from `LocationManager` class. When a registered callback method is called, it is provided with an input parameter referring to a `Location` object. The creator of this object is designated to the application in the current sandbox by the object's created cell. Furthermore, it is marked as sensitive in its sensitive cell (see Fig. 10).

The module Intent is responsible for implementing the semantics of API methods for creating and manipulating intents such as the constructor of `Intent` class, adding extra string data into an intent, i.e., `putExtra` from `Intent` class and retrieving extra string data from an intent, i.e., `getStringExtra` method also from `Intent` class. The module Broadcast is responsible for implementing the semantics of API methods relating to broadcasting intents, for example: broadcasting an intent, i.e., `sendBroadcast` method from `Context` class; and registering an callback function when receiving an broadcasted intent, i.e., `registerReceiver` method from `Context` class. In particular, when executing `sendBroadcast` with an intent, this intent will be placed in the `broadcasts` cell (see Fig. 9) in the configuration. Then, callback methods previously registered by a call to `registerReceiver` will be called according to the newly placed intent in `broadcasts` cell. Finally, the module `Apache-Http` implements the semantics of methods relating to sending http request, i.e. `execute` method from `HttpUriRequest` class.

5.3.4 Detecting Collusion on the Concrete Semantics Level

Finally, we detect information theft via collusion by annotating any "object" cell with two additional values: "sensitive" and "created". *Sensitive* is a Boolean value indicating if the object is sensitive (e.g., device locations, contacts, private data, etc.). *Created* is a set of app ids that initialise the object. Information theft collusion is conducted when one app collects sensitive data from an Android[TM] device and forwards to another app who will export it outside the device boundaries. In detail, this process includes, within a device: (1) a sensitive object O_1 is initialised by an app A_1, i.e., *Sensitive* of O_1 is *true* and *Created* of O_1 contains id A_1; (2) O_1 (or its content) is forwarded to another app A_2 via communication (possibly through a series of actions of creating new objects or editing existing objects using O_1 where their *Sensitive* is updated to that of O_1 and their *Created* is updated to include A_1); (3) A_2 calls an API to export O_1 (or any of these objects whose *Sensitive* is true and *Created* contains id $A_1 \neq A_2$). Information theft collusion is detected in our framework when A_2 calls the API to export an object with *Sensitive* equal true and *Created* containing any id $A_1 \neq A_2$.

This characterisation of collusion as an information flow property implies the conditions of Definition 1:

- A_1 contributes in retrieving the sensitive data while A_2 is exporting.
- A_1 and A_2 communicate with each other to transfer the sensitive data from A_1 to A_2.

5.4   An Abstract Semantics for Dalvik

The abstract semantics lightens the configuration and the transitions in order to gain efficiency for model checking while maintaining enough information to verify collusion. The abstract configuration has a cell structure similar to the concrete configuration except for the memory cell: instead of creating objects, in abstract semantics we record the information flow by propagating the object types and the constants (either strings or numerical). Structurally, the $\mathbb{K}$ specification for the abstract semantics is organised in the same way as the concrete one, c.f. Fig. 12. In the followings we describe the differences that render the abstraction.

In the "read/write" module the abstract semantics neglects the memory-related details as described next: The abstract semantics for the *instructions that create new object instances* (e.g., "new-instance *Register, Type*") sets the register to the type of the newly created object. The *arithmetic instructions* only define data dependency between the source registers and the destination register. The *move instruction*, that copies one register into another, sets the contents of the source register into the destination. Finally, the *load/store instructions*, that copy from or into the memory, are similarly abstracted into data-dependence. We exemplify this latest class of instructions with the abstract semantics of the iget instruction in Fig. 14.

The abstract semantics is field insensitive, e.g., the iget instruction only maintains the information collected in the object register, R_2. In order to add field sensitivity to the abstraction, we only need to en-queue in R_1 the field F such that after the substitution we have $R_1 \mapsto F \curvearrowright L2$.

The module "invoke/return" contains the most significant differences of the abstract semantics w.r.t. the concrete semantics. The *invoke instructions* differentiate the API calls from the app's methods. The methods defined in the app are executed upon invocation (using a call stack) while the API calls are further discriminated into app-communication (i.e., "send" or "receive"), APIs which trigger callbacks, APIs which access sensitive data, APIs which publish data, and ordinary APIs. We currently consider only Intent based inter-app communication. All invoke instructions add information to the data-flow as follows: the object for which the method is invoked depends on the parameters of the invoked method. Similarly, the *move-result instruction* defines data-dependence between the parameters of the latest invoked method and the register where the result is written. The data-flow abstraction allows us to see an API call just as an instruction producing additional data dependency. Hence, we do not need to treat separately these APIs as in the

Fig. 14 $\mathbb{K}$ rule for the abstract semantics of iget instruction

concrete semantics (by either executing their code or giving them special semantics). This gives a lightweight feature to the abstract semantics meanwhile enlarging the class of apps that can be verified. Obviously, the price paid is the over approximation of the app behaviours which induces false positive colluding results.

The rules producing transitions in the transition system are defined in the "control" module. The rules for *branching instructions*, i.e., if-then instructions, are always considered non-deterministic in the abstract semantics. The rules for goto instruction check if the goto destination was already traversed in the current execution and, if this is the case, the jump to the destination is replaced by a fall through. As such, the loops are traversed at most once since the data-flow collection only requires one loop traversal.

5.4.1 Detecting Collusion on the Abstract Semantics Level

Detecting collusion on the abstract semantics level works as follows: When an API accessing sensitive data is invoked in an app A_1, the data-flow is augmented with special a label "*secret*(A_1)". If, via the data-flow abstraction, the "secret" arrives into the parameters of a publish invocation of a different app A_2 ($A_1 \neq A_2$) then we discover a collusion pattern for information theft. Note that the "secret" could be passed from A_1 to A_2 directly or via other apps $A's$.

The property detected in the abstract semantics is a safe over-approximation of Definition 1. Namely, (1) the set of colluding apps S includes two different apps A_1 and A_2, hence S is not a singleton set; (2) the apps A_1 and A_2 execute the beginning and the end of the threat (i.e. the extraction and the publication of the secret, respectively) while the apps $A's$ act as messengers; (3) all the discovered apps contribute in communicating the secret.

Note, we say that the abstract collusion result is an over-approximation due to the fact that only "non-colluding" results could be a guarantee for the non-existence of a set S with the characteristics given by Definition 1. If a colluding set S is reported in the abstract model checking then this is either a true collusion, as argued in (1–3), or a false witness to collusion. A false witness (also named "spurious counterexample" in abstract model checking) may appear due to the overprotective nature of the data-flow abstraction. This abstraction assumes that any data "touching" the secret may take it and pass it (e.g. when the secret is given as parameter to an API call f then any other parameter and the result of f are assumed to know the secret). Consequently, any collusion set S reported by the abstract model checking has to be verified (e.g. by exercising the concrete semantics over S).

5.5 Experimental Results

We demonstrate how collusion is detected using our concrete and our abstract semantics on two Android™ applications, called LocSender and LocReceiver. Together, these two apps jointly carry out an "information theft".

They consist of about 100 lines of Java code/3000 lines of Smali code each. Originally written to explore if collusion was actually possible (there is no APK of the Soundcomber example), here they serve as a test for our model checking approach.

LocSender obtains the location of the Android™ device and communicates it using a broadcast intent. LocReceiver constantly waits for such a broadcast. On receiving such message, it extracts the location information and finally sends it to the Internet as an HTTP request. We have two variants of LocReceiver: one contains a while loop pre-processing the HTTP request while the other does not. Additionally, we create two further versions of each LocReceiver variant where collusion is broken by (1) not sending the HTTP request at the end, (2) altering the name of the intent that it waits for—named LocReceiver1 and LocReceiver2, respectively. Furthermore, we (3) create a LocSender1 which sends a non-sensitive piece of information rather than the location. In total, we will have eight experiments where the two firsts have a collusion while the six lasts do not.[9] Figure 15 summarises the experimental results.

5.5.1 Evaluation

Our experiments indicate that our approach works correctly: if there is collusion it is either detected or has a timeout, if there is no collusion then none is detected. In case of detection, we obtain a trace providing evidence of a run leading to information theft. The experiments further demonstrate the need for an abstract semantics,

App1	App2	Loop	Collusion	Concrete		Abstract	
				Runtime	Detected	Runtime	Detected
LocSender	LocReceiver		✓	55s	✓	30s	✓
LocSender	LocReceiver	✓	✓	time-out		33s	✓
LocSender	LocReceiver1			1m13s		31.984s	
LocSender	LocReceiver1	✓		time-out		34s	
LocSender	LocReceiver2			53s		32s	
LocSender	LocReceiver2	✓		time-out		33s	
LocSender1	LocReceiver			1m11s		32s	
LocSender1	LocReceiver	✓		time-out		34s	

Fig. 15 Experimental result

---

[9]All experiments are carried out on a Macbook Pro with an Intel i7 2.2 GHz quad-core processor and 16 GB of memory.

beyond the obvious argument of speed: e.g. in case of a loop where the number of iterations depends on an environmental parameter that can't be determined, the concrete semantics yields a time out, while the abstract semantics still is able to produce a result. Model checking with the abstract semantics is about twice as fast as with the concrete semantics. At least for such small examples, our approach appears to be feasible.

6   Related Work

In this section we review the different previous works that have addressed the identification and prevention of Android[TM] malicious software. We first review previous approaches to detect and identify Android[TM] malware (single apps) in general. Then, we address previous work on detection and identification of colluding apps. Finally, we review works that focus on collusion prevention.

6.1   Detecting Malicious Applications

In general, techniques for detecting Android[TM] malware are categorised into two groups: static and dynamic. In static analysis, certain features of an app are extracted and analysed using different approaches such as machine learning techniques. For example, Kirin [26] proposes a set of policies which allows matching permissions requested by an app as an indication for potentially malicious behaviour. DREBIN [4] trained Support Vector Machines for classifying malware using number of features: used hardware components, requested permissions, critical and suspicious API calls and network addresses. Similar static techniques can be found in [16, 19, 38, 44, 63]. Conversely, dynamic analysis detects malware at run-time. It deploys suitable monitors on Android[TM] systems and constantly looks for malicious behaviours imposed by software within the system. For example, [33] keeps track of the network traffic (DNS and HTTP requests in particular) in an Android[TM] system as input and then utilises Naive Bayes Classifier in order to detect malicious behaviours. Similarly, [39] collects information about the usage of network (data sent and received), memory and CPU and then uses multivariate time-series techniques to decide if an app admitted malicious behaviours. A different approach is to translate Android[TM] apps into formal specifications and then to employ the existing model checking techniques. These explore all possible runs of the apps in order to search for a matching malicious activity represented by formulas of some temporal logic, see, e.g., [10, 60].

In contrast to malware detection, detecting colluding apps involves not only identifying whether a security threat can be carried out by these apps but also revealing whether communication between them occurs during the attack. In other words, existing malware detection techniques are not directly applicable for detecting collusion.

6.2 Detecting Malicious Inter-app Communication

Current research mostly focuses on detecting inter-app communication and information leakage. DidFail [15] is a analysis tool for Android™ apps that detects possible information flows between multiple apps. Each APK is fed into the APK transformer, a tool that annotates intent-related function calls with information that uniquely identifies individual cases where intents are used in the app, and then the transformed APK is passed to two other tools: FlowDroid [5, 28] and Epicc [49]. The FlowDroid tool performs static taint tracking in Android™ apps. That analysis is field, flow and context sensitive with some object sensitivity. Epicc performs static analysis to map out inter-component communication within an Android™ app. Epicc [49] provides flow and context sensitive analysis for app communication, but it does not tackle each and every possible communication channels between apps' components. The most similar work to DidFail is IccTA [42] which statically analyses app sets to detect flows of sensitive data. IccTA uses a single-phase approach that runs the full analysis monolithically, as opposed to DidFail's composition two-phase analysis. DidFail authors acknowledge the fact that IccTA is more precise than the current version of DidFail because of its greater context sensitivity. This supports our claim in Sect. 4.2—"context would be the key" for improving the precision. FUSE [52], a static information flow analysis tool for multi-apps, provides similar functions as DidFail and IccTA in addition to visualising inter-component communication (ICC) maps. DroidSafe [31] is a static information flow analysis tool to report potential leaks of sensitive information in Android™ applications.

ComDroid [17] detects app communication vulnerabilities. Automatic detection of inter-app permission leakage is provided [56]. Authors address three kinds of such attacks: confused deputy, permission collusion and intent spoofing and use taint analysis to detect them. An empirical evaluation of the robustness of ICC through fuzz testing can be found in [45]. A study of network covert channels on Android™ is [29, 30]. Authors show that covert channels can be successfully implemented in Android™ for data leakage. A security framework for Android™ to protect against confused deputy and collusion attacks is proposed [14]. The master thesis [53] provides an analysis of covert channels on mobile devices. COVERT [9] is a tool for compositional analysing inter-app vulnerabilities. TaintDroid [25], an information-flow tracking system, provides a real time analysis by leveraging Android™'s virtualized execution environment. DroidForce [51], build upon on FlowDroid, attempts to addresses app collusion problem with a dynamic enforcement mechanism backed by a flexible policy language. However static analysis encourages in collusion detection due the scalability and completeness issues [24]. Desired properties for a practical solution include, but are not limited to: characterising the context associated with communication channels with fine granularity, minimising false alarms and ability to scalable for a large number of apps.

6.3   Other Approaches

Application collusion can also be mitigated by implementing compartmentalisation techniques, like Samsung Knox® [1]. These techniques isolate app groups by forbidding any communication between apps in different groups. In [61] the authors analyse several compartmentalisation strategies to minimise the risk of app collusion. Their results show that it is enough to have two or three app compartments to greatly reduce the risk posed by a set of 20–50 apps. In order to reduce the risk further, the amount of app compartments must be increased exponentially.

Finally, Bartel et al. [43] propose the tool *APKCombiner* which joins two apps into a single APK file. In this way, a security analyst can use inter-component, instead of inter-app, communication analysers to analyse the inter app communication mechanisms that exist between apps. From an evaluation set of 3000 apps they were able of joining 88% of them. The average time required to join two apps with APKCombiner is 3 min. This makes it hard to use for practical app analysis.

7   Conclusion and Future Work

We summarise the state of the art w.r.t. collusion prevention and point the reader to the current open research questions in the field.

A frontal approach to detecting collusions to analyse pairs, triplets and larger sets is not practical given the search space. Thus, we consider the step of pre-filtering apps essential for a collusion detection system if it were to be used in practice. Even if we could find all collusions in all existing apps, new ones appear every day and they could create new collusions with previously analysed apps. Continuously re-analysing the growing space of all Android™ apps is unfeasible so an effective collusion-discovery tool must include an effective set of methods to isolate potential sets which require further examination.

The best long-term solution would be to enforce more isolation in the Android™ OS itself. For example, apps may be required to explicitly declare all communications (this includes not only inter-app channels but also declaring all Internet domains, ports and services which they intend to use) via their manifests and then the OS will be able to block all other undeclared communications. However, this will not work for already existing apps (as well as many apps which could be created before such OS hardening were implemented) so in the meantime the best practical approach is to employ, enhance and expand the array of filtering mechanisms we developed to discover potentially colluding sets of apps.

A filter based on Android™ app permissions is the simplest one. Permissions are very easy and cheap to extract from APKs—no de-compilation, reverse engineering, complex code or data flow analysis is required. Alternatively (or additionally), to the two filters described in our chapter, imprecise heuristic methods to find "interesting" app sets may include: statistical code analysis of apps (e.g. to locate APIs potentially responsible to communication, accessing sensitive information, etc.); and taking into account apps' publication time and distribution channel (app market, direct installation, etc.).

Attackers are more likely to release colluding apps in a relatively short time frame and they are likely to engineer the distribution in such a way that a sufficient number of users would install the whole set (likely from the same app market). To discover such scenarios one can employ: analysis of security telemetry focused on users' devices to examine installation/removal of apps, list of processes simultaneously executing, device-specific APK download/installation logs from app markets (like Google Play™) and meta-data about APKs in app markets (upload time by developers, developer ID, source IP, etc.). Such data would allow constructing a full view of existing app sets on user devices. Only naturally occurring sets (either installed on same device or actually executing simultaneously) may be analysed for collusion which should drastically reduce the number of sets that require deeper analysis.

Naturally, finding "interesting" app sets is not enough: in the end, some analysis is required to figure out if a given set of apps colludes. Manual analysis is costly, merging apps into a single one often fails, however software model checking of suitable abstractions of an app set might be a way forward. We demonstrated that both semantic approaches are—in principle—able to successfully model check for app collusion realising the threat of information theft. Here, naturally the abstract semantics outperforms the concrete one. Though it is still early days, we dare to express the following expectation: we believe that our approach will scale thanks to its powerful built-in abstraction mechanisms.

The aspiration set out in this chapter is to build a fully automated and effective collusion detection system, and tool performance will be central to address scale. It is not clear yet where the bottleneck will be when we apply our approach to real-life apps in a fully operational deployment. Further work will focus on identifying these bottlenecks to optimise the slowest elements of our tool-chain. Detecting covert channels would be a challenge as modelling such will not be trivial; this is the natural next step.

In the long run, collusions are a part of a general problem of effective isolation of software. This problem exists in all environments which implement sandboxing of software—from other mobile operating systems (like iOS® and Tizen®) to virtual machines in server farms (like Amazon EC2, Microsoft Azure and similar). We can see how covert communications between sandboxes may be used to breach security and create data leaks. The tendency to have more and better isolation is, of course, a positive one but we should fully expect the attackers to employ collusion methods more often to circumvent security. We endeavour to see if our methods developed for Android™ would be applicable to a wider range of operating environments.

Acknowledgements This work has been supported by UK Engineering and Physical Sciences Research Council (EPSRC) grant EP/L022699/1. The authors would like to thank the anonymous reviewers for their helpful comments, and Erwin R. Catesbeiana (Jr) for pointing out the importance of intention in malware analysis.

References

1. (2016). URL https://www.samsungknox.com/
2. Android^TM Package Index. http://developer.android.com/reference/packages.html (2016)
3. Android^TM Open Source Project: Dalvik Bytecode. https://source.android.com/devices/tech/dalvik/dalvik-bytecode.html (2016)
4. Arp, D., Spreitzenbarth, M., Hubner, M., Gascon, H., Rieck, K.: DREBIN: effective and explainable detection of Android^TM malware in your pocket. In: 21st Annual Network and Distributed System Security Symposium, NDSS 2014, San Diego, California, USA, February 23–26, 2014. The Internet Society (2014). URL http://www.internetsociety.org/doc/drebin-effective-and-explainable-detection-android-malware-your-pocket
5. Arzt, S., Rasthofer, S., Fritz, C., Bodden, E., Bartel, A., Klein, J., Le Traon, Y., Octeau, D., McDaniel, P.: FlowDroid: precise context, flow, field, object-sensitive and lifecycle-aware taint analysis for Android^TM apps. In: ACM SIGPLAN Notices - PLDI'14, vol. 49, pp. 259–269. ACM (2014)
6. Asavoae, I.M., Blasco, J., Chen, T.M., Kalutarage, H.K., Muttik, I., Nguyen, H.N., Roggenbach, M., Shaikh, S.A.: Towards Automated Android^TM App Collusion Detection. CoRR **abs/1603.02308** (2016). URL http://arxiv.org/abs/1603.02308
7. Asavoae, I.M., Muttik, I., Roggenbach, M.: Android^TM malware: They divide, we conquer. Bucharest, Romania (2016)
8. Asavoae, I.M., Nguyen, H.N., Roggenbach, M., Shaikh, S.A.: Utilising $\mathbb{K}$ Semantics for Collusion Detection in Android^TM Applications. In: Critical Systems: Formal Methods and Automated Verification - Joint 21st International Workshop on Formal Methods for Industrial Critical Systems and 16th International Workshop on Automated Verification of Critical Systems, FMICS-AVoCS 2016, Pisa, Italy, September 26–28, 2016, Proceedings, pp. 142–149 (2016). DOI 10.1007/978-3-319-45943-$1_1$0
9. Bagheri, H., Sadeghi, A., Garcia, J., Malek, S.: Covert: Compositional analysis of Android^TM inter-app vulnerabilities. Tech. rep., Tech. Rep. GMU-CS-TR-2015-1, Department of Computer Science, George Mason University, 4400 University Drive MSN 4A5, Fairfax, VA 22030-4444 USA (2015)
10. Beaucamps, P., Gnaedig, I., Marion, J.: Abstraction-based malware analysis using rewriting and model checking. In: S. Foresti, M. Yung, F. Martinelli (eds.) Computer Security - ESORICS 2012 - 17th European Symposium on Research in Computer Security, Pisa, Italy, September 10–12, 2012. Proceedings, *Lecture Notes in Computer Science*, vol. 7459, pp. 806–823. Springer (2012). DOI 10.1007/978-3-642-33167-1_46. URL http://dx.doi.org/10.1007/978-3-642-33167-1_46
11. Blasco, J., Chen, T., Muttik, I., Roggenbach, M.: Wild Android^TM collusions. Virus Bulletin 2016 (2016)
12. Blasco, J., Chen, T.M., Muttik, I., Roggenbach, M.: Efficient Detection of App Collusion Potential Using Logic Programming. IEEE Transactions on Mobile Computing (2017). arXiv:1706.02387. http://arxiv.org/abs/1706.02387
13. Blasco, J., Muttik, I.: Android^TM collusion conspiracy (2015)
14. Bugiel, S., Davi, L., Dmitrienko, A., Fischer, T., Sadeghi, A.R., Shastry, B.: Towards taming privilege-escalation attacks on Android^TM. In: NDSS (2012)
15. Burket, J., Flynn, L., Klieber, W., Lim, J., Snavely, W.: Making didfail succeed: Enhancing the cert static taint analyzer for Android^TM app sets. Tech. Rep. MSU-CSE-00-2, Software Engineering Institute, Carnegie Mellon University, Pittsburgh,USA (2015)
16. Canfora, G., Lorenzo, A.D., Medvet, E., Mercaldo, F., Visaggio, C.A.: Effectiveness of opcode ngrams for detection of multi family Android^TM malware. In: 10th International Conference on Availability, Reliability and Security, ARES 2015, Toulouse, France, August 24–27, 2015, pp. 333–340 (2015). DOI 10.1109/ARES.2015.57. URL http://dx.doi.org/10.1109/ARES.2015.57

17. Chin, E., Felt, A.P., Greenwood, K., Wagner, D.: Analyzing inter-application communication in Android™. In: MobiSys'11, pp. 239–252 (2011)
18. Clavel, M., Duran, F., Eker, S., Lincoln, P., Martı-Oliet, N., Meseguer, J., Talcott, C.: All about Maude. LNCS **4350** (2007)
19. Dai, G., Ge, J., Cai, M., Xu, D., Li, W.: SVM-based malware detection for Android™ applications. In: Proceedings of the 8th ACM Conference on Security & Privacy in Wireless and Mobile Networks, New York, NY, USA, June 22–26, 2015, pp. 33:1–33:2 (2015). DOI 10.1145/2766498.2774991. URL http://doi.acm.org/10.1145/2766498.2774991
20. Desnos, A.: Androguard. https://github.com/androguard/androguard (2016)
21. Dubey, A., Misra, A.: Android™ Security: Attacks and Defenses. CRC Press (2013)
22. Elenkov, K.: Android™ Security Internals: An In-Depth Guide to Android™'s Security Architecture. No Starch Press (2014)
23. Elish, K.O., Yao, D., Ryder, B.G.: On the need of precise inter-app ICC classification for detecting Android™ malware collusions. In: MoST (2015)
24. Elish, K.O., Yao, D.D., Ryder, B.G.: On the need of precise inter-app icc classification for detecting Android™ malware collusions. In: Proceedings of IEEE Mobile Security Technologies (MoST), in conjunction with the IEEE Symposium on Security and Privacy (2015)
25. Enck, W., Gilbert, P., Han, S., Tendulkar, V., Chun, B.G., Cox, L.P., Jung, J., McDaniel, P., Sheth, A.N.: Taintdroid: an information-flow tracking system for realtime privacy monitoring on smartphones. ACM Transactions on Computer Systems (TOCS) **32**(2), 5 (2014)
26. Enck, W., Ongtang, M., McDaniel, P.: On lightweight mobile phone application certification. In: Proceedings of the 16th ACM conference on Computer and communications security, pp. 235–245. ACM (2009)
27. Enck, W., Ongtang, M., McDaniel, P.: Understanding Android™ security. IEEE security & privacy (1), 50–57 (2009)
28. Fritz, C., Arzt, S., Rasthofer, S., Bodden, E., Bartel, A., Klein, J., le Traon, Y., Octeau, D., McDaniel, P.: Highly precise taint analysis for Android™ applications. EC SPRIDE, TU Darmstadt, Tech. Rep (2013)
29. Gasior, W., Yang, L.: Network covert channels on the Android™ platform. In: Proceedings of the Seventh Annual Workshop on Cyber Security and Information Intelligence Research, p. 61. ACM (2011)
30. Gasior, W., Yang, L.: Exploring covert channel in Android™ platform. In: Cyber Security (CyberSecurity), 2012 International Conference on, pp. 173–177 (2012). DOI 10.1109/Cyber-Security.2012.29
31. Gordon, M.I., Kim, D., Perkins, J.H., Gilham, L., Nguyen, N., Rinard, M.C.: Information flow analysis of Android™ applications in droidsafe. In: NDSS (2015)
32. Gunasekera, S.: Android™ Apps Security. Apress (2012)
33. Han, H., Chen, Z., Yan, Q., Peng, L., Zhang, L.: A real-time Android™ malware detection system based on network traffic analysis. In: Algorithms and Architectures for Parallel Processing - 15th International Conference, ICA3PP 2015, Zhangjiajie, China, November 18–20, 2015. Proceedings, Part III, pp. 504–516 (2015). DOI 10.1007/978-3-319-27137-8_37. URL http://dx.doi.org/10.1007/978-3-319-27137-8_37
34. Hardy, N.: The confused deputy:(or why capabilities might have been invented). ACM SIGOPS Operating Systems Review **22**(4), 36–38 (1988)
35. Harley, D., Lee, A.: Antimalware evaluation and testing. In: AVIEN Malware Defense Guide. Elsevier (2007)
36. Huskamp, J.C.: Covert communication channels in timesharing systems. Ph.D. thesis, California Univ., Berkeley (1978)
37. Kalutarage, H.K., Nguyen, H.N., Shaikh, S.A.: Towards a threat assessment for apps collusion. Telecommunication Systems 1-14 (2016). doi:10.1007/s11235-017-0296-1. http://dx.doi.org/10.1007/s11235-017-0269-1

38. Kate, P.M., Dhavale, S.V.: Two phase static analysis technique for Android^{TM} malware detection. In: Proceedings of the Third International Symposium on Women in Computing and Informatics, WCI 2015, co-located with ICACCI 2015, Kochi, India, August 10–13, 2015, pp. 650–655 (2015). DOI 10.1145/2791405.2791558. URL http://doi.acm.org/10.1145/2791405.2791558

39. Kim, K., Choi, M.: Android^{TM} malware detection using multivariate time-series technique. In: 17th Asia-Pacific Network Operations and Management Symposium, APNOMS 2015, Busan, South Korea, August 19–21, 2015, pp. 198–202 (2015). DOI 10.1109/APNOMS.2015.7275426. URL http://dx.doi.org/10.1109/APNOMS.2015.7275426

40. Klieber, W., Flynn, L., Bhosale, A., Jia, L., Bauer, L.: Android^{TM} taint flow analysis for app sets. In: Proceedings of the 3rd ACM SIGPLAN International Workshop on the State of the Art in Java Program Analysis, pp. 1–6. ACM (2014)

41. Krishnamoorthy, K.: Handbook of statistical distributions with applications. CRC Press (2015)

42. Li, L., Bartel, A., Bissyand, T., Klein, J., Le Traon, Y., Arzt, S., Siegfried, R., Bodden, E., Octeau, D., Mcdaniel, P.: IccTA: Detecting Inter-Component Privacy Leaks in Android^{TM} Apps. In: Proceedings of the 37th International Conference on Software Engineering (ICSE 2015) (2015)

43. Li, L., Bartel, A., Bissyandé, T.F., Klein, J., Le Traon, Y.: ApkCombiner: Combining multiple Android^{TM} apps to support inter-app analysis. In: SEC'15, pp. 513–527. Springer (2015)

44. Li, Q., Li, X.: Android^{TM} malware detection based on static analysis of characteristic tree. In: 2015 International Conference on Cyber-Enabled Distributed Computing and Knowledge Discovery, CyberC 2015, Xi'an, China, September 17–19, 2015, pp. 84–91 (2015). DOI 10.1109/CyberC.2015.88. URL http://dx.doi.org/10.1109/CyberC.2015.88

45. Maji, A.K., Arshad, F., Bagchi, S., Rellermeyer, J.S., et al.: An empirical study of the robustness of inter-component communication in Android^{TM}. In: Dependable Systems and Networks (DSN), 2012 42nd Annual IEEE/IFIP International Conference on, pp. 1–12. IEEE (2012)

46. Marforio, C., Francillon, A., Capkun, S.: Application collusion attack on the permission-based security model and its implications for modern smartphone systems. technical report (2011)

47. Marforio, C., Ritzdorf, H., Francillon, A., Capkun, S.: Analysis of the communication between colluding applications on modern smartphones. In: Proceedings of the 28th Annual Computer Security Applications Conference, pp. 51–60. ACM (2012)

48. Muttik, I.: Partners in crime: Investigating mobile app collusion. In: McAfee® Threat Report (2016)

49. Octeau, D., McDaniel, P., Jha, S., Bartel, A., Bodden, E., Klein, J., Le Traon, Y.: Effective inter-component communication mapping in Android^{TM} with epicc: An essential step towards holistic security analysis. In: USENIX Security 2013 (2013)

50. Peng, H., Gates, C., Sarma, B., Li, N., Qi, Y., Potharaju, R., Nita-Rotaru, C., Molloy, I.: Using probabilistic generative models for ranking risks of Android^{TM} apps. In: Proceedings of the 2012 ACM conference on Computer and communications security, pp. 241–252. ACM (2012)

51. Rasthofer, S., Arzt, S., Lovat, E., Bodden, E.: Droidforce: enforcing complex, data-centric, system-wide policies in Android^{TM}. In: Availability, Reliability and Security (ARES), 2014 Ninth International Conference on, pp. 40–49. IEEE (2014)

52. Ravitch, T., Creswick, E.R., Tomb, A., Foltzer, A., Elliott, T., Casburn, L.: Multi-app security analysis with fuse: Statically detecting Android^{TM} app collusion. In: Proceedings of the 4th Program Protection and Reverse Engineering Workshop, p. 4. ACM (2014)

53. Ritzdorf, H.: Analyzing covert channels on mobile devices. Ph.D. thesis, ETH Zürich, Department of Computer Science (2012)

54. Roşu, G., Şerbănuţă, T.F.: An overview of the K semantic framework. Journal of Logic and Algebraic Programming **79**(6), 397–434 (2010)

55. Sarma, B.P., Li, N., Gates, C., Potharaju, R., Nita-Rotaru, C., Molloy, I.: Android^{TM} permissions: a perspective combining risks and benefits. In: Proceedings of the 17th ACM symposium on Access Control Models and Technologies, pp. 13–22. ACM (2012)

56. Sbirlea, D., Burke, M., Guarnieri, S., Pistoia, M., Sarkar, V.: Automatic detection of inter-application permission leaks in Android™ applications. IBM Journal of Research and Development **57**(6), 10:1–10:12 (2013). DOI 10.1147/JRD.2013.2284403

57. Schlegel, R., Zhang, K., Zhou, X.y., Intwala, M., Kapadia, A., Wang, X.: Soundcomber: A stealthy and context-aware sound trojan for smartphones. In: NDSS'11, pp. 17–33 (2011)

58. Shen, S.: Setting the record straight on moplus sdk and the wormhole vulnerability. http://blog.trendmicro.com/trendlabs-security-intelligence/setting-the-record-straight-on-moplus-sdk-and-the-wormhole-vulnerability/. Accessed: 04/0/2016

59. Six, J.: Application Security for the Android™ Platform: Processes, Permissions, and Other Safeguards. O'Reilly (2011)

60. Song, F., Touili, T.: Model-checking for Android™ malware detection. In: J. Garrigue (ed.) Programming Languages and Systems - 12th Asian Symposium, APLAS 2014, Singapore, November 17–19, 2014, Proceedings, *Lecture Notes in Computer Science*, vol. 8858, pp. 216–235. Springer (2014). DOI 10.1007/978-3-319-12736-1_12. URL http://dx.doi.org/10.1007/978-3-319-12736-1_12

61. Suarez-Tangil, G., Tapiador, J.E., Peris-Lopez, P.: Compartmentation policies for Android™ apps: A combinatorial optimization approach. In: Network and System Security, pp. 63–77 (2015)

62. Suarez-Tangil, G., Tapiador, J.E., Peris-Lopez, P., Ribagorda, A.: Evolution, detection and analysis of malware for smart devices. Comm. Surveys & Tutorials, IEEE **16**(2), 961–987 (2014)

63. Wang, Z., Li, C., Guan, Y., Xue, Y.: Droidchain: A novel malware detection method for Android™ based on behavior chain. In: 2015 IEEE Conference on Communications and Network Security, CNS 2015, Florence, Italy, September 28–30, 2015, 727–728 (2015). DOI 10.1109/CNS.2015.7346906. URL http://dx.doi.org/10.1109/CNS.2015.7346906

Dynamic Analysis of Malware Using Run-Time Opcodes

Domhnall Carlin, Philip O'Kane, and Sakir Sezer

Abstract The continuing fight against intentionally malicious software has, to date, favoured the proliferators of malware. Signature detection methods are growingly impotent against rapidly evolving obfuscation techniques. Research has recently focussed on the low-level opcode analysis of disassembled executable programs, both statically and dynamically. While able to detect malware, static analysis often still cannot unravel obfuscated code; dynamic approaches allow investigators to reveal the run-time code. Old and inadequately sampled datasets have limited the extrapolation potential of much of the body of research. This work presents a dynamic opcode analysis approach to malware detection, applying machine learning techniques to the largest dataset of its kind, both in terms of breadth (610–100k features) and depth (48k samples). N-gram analysis of opcode sequences from $n = 1..3$ was applied as a means of enhancing the feature set. Feature selection was then investigated to tackle the feature explosion which resulted in more than 100,000 features in some cases. As the earliest detection of malware is the most favourable, run-length, i.e. the number of recorded opcodes in a trace, was examined to find the optimal capture size. This research found that dynamic opcode analysis can detect malware from benignware with a 99.01% accuracy rate, using a sequence of only 32k opcodes and 50 features. This demonstrates that a dynamic opcode analysis approach can compare with static analysis in terms of speed. Furthermore, it has a very real potential application to the unending fight against malware, which is, by definition, continuously on the back foot.

D. Carlin (✉) • P. O'Kane • S. Sezer
Centre for Secure Information Technologies, Queen's University, Belfast, Northern Ireland, UK
e-mail: dcarlin05@qub.ac.uk; p.okane@qub.ac.uk

© Springer International Publishing AG 2017
I. Palomares et al. (eds.), *Data Analytics and Decision Support for Cybersecurity*,
Data Analytics, DOI 10.1007/978-3-319-59439-2_4

99

1 Introduction

Since the 1986 release of *Brain*, regarded as the first PC virus, both the incessant proliferation of malware (**mal**icious soft**ware**) and the failure of standard anti-virus (AV) software, have posed significant challenges for all technology users. In the second quarter of 2016, McAfee Labs reported that the level of previously unseen instances of malware had exceeded 40 million for the preceding three months. The total number of *unique* malware samples reached 600 million, a growth of almost one-third on the previous year [4].

Malware was previously seen as the product of miscreant teenagers, whereas it is now largely understood to be a highly effective tool for criminal gangs and alleged national malfeasance. As a tool, malware facilitates a worldwide criminal industry and enables theft and extortion on a global scale. One data breach in a US-based company has been estimated to cost an average $5.85 million, or $102 for each record compromised [1]. GDP estimates which attribute the cost of cybercrime to modern developed countries are as much as 1.6% in Germany, with other technology-heavy nations showing alarming levels (USA—0.64% and China—0.63%) [4].

This chapter is presented in the following way: the remainder of the current section describes malware, techniques for detecting malicious code and subsequent counter-attacks. Section 2 presents related research in the field and the gap in the current body of literature. Section 3 depicts the methodology used in establishing the dataset and Section 4 presents the results of machine learning analyses of the data. The final section discusses the conclusions which can be reached about the results presented, and describes potential future work.

1.1 Taxonomy of Malware

Malware can be categorised according to a variety of features, but typically according to its family type. Bontchev, Skulason and Solomon in [8] established naming conventions for malware, which gave some consideration to the taxonomy of such files. With the rapid advancement of malware, this original naming convention required updating to cope with the new standards in developing malware [15]. The standard name of a unique piece of malware is generally comprised of four parts, as seen in Fig. 1, although the format presented (delimiters etc) varies widely according to the organisation.

The pertinent parts of the name are *Type* and *Family*, as these provide the broad delineations between categories. In the original naming standard [8], noted that it was virtually impossible to adequately formally define a malware family, but that structurally similar malware variants could be grouped as a family e.g. MyDoom, Sasser etc. Malware is categorised according to *Type*, when the behaviour of the malware and resultant effect on the victim system is considered.

> *Type.Platform.Family-Name.Minor-Variant : Modifier*
>
> *e.g.Trojan.Win32.Crilock.B :*
>
> = *A Windows-based Trojan, and the second variant found from the Crilock family*

Fig. 1 Naming convention for malware

Such types include:

- *Backdoor*: installed onto the victim machine, allowing the attacker remote access;
- *Botnet*: as per a *back door*, but the victim machine is one of a cluster of breached machines linked to a command-and-control server;
- *Downloader*: preliminary code which serves to download further malicious code;
- *Information-stealing malware*: code which scrapes information from the victim machine and forwards it to an intended source e.g. banking info;
- *Launcher*: used to launch other malicious programs through non-legitimate methods in order to maintain stealth;
- *Rootkit*: allows access to unauthorised areas of a machine and designed to conceal the existence of other malcode;
- *Scareware*: often imitates an anti-virus (AV) alert, designed to scare an uninformed user into purchasing other software;
- *Spam-sending malware*: causes the victim machine to send spam unknowingly;
- *Virus*: code that can copy itself and infect additional computers, but requires a host file;
- *Worm*: malware which self-propagates via network protocols;
- *Trojan*: malware which is normally dressed up as innocent software and does not spread by itself, but allows access to the victim machine;
- *Ransomware*: malware which encrypts the victim machine's files, extorting money from the victim user for their release;
- *Dropper*: designed to bypass AV scanners and contains other malware which is unleashed from within the original code, after the installation of the original Dropper.

As malware very often overlaps categories, or implements several mechanisms, a clear taxonomy system can be a difficult task [25]. Deferring then to *family* as the discrete nominal attribute can provide an adequate definition of the malware being referenced. For example, the well-publicised and highly-advanced *Stuxnet* cyberweapon, which was used to attack and disrupt the Iranian nuclear program, spans multiple categories of malware. The initial attack vector was through an infected USB drive, which contains the worm module and a *.lnk* file to the worm itself. Stuxnet exploits several vulnerabilities, one of which allows automatic execution of *.lnk* files on USB drives. The worm module contains the routines for the payload and a third module, a rootkit, hides the malicious activities and codes from the end-user [42]. As this is a multifaceted attack, it could be difficult to adequately describe the

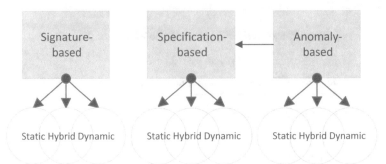

Fig. 2 Malware detection methods

type of the malware using a traditional approach. However, referring to the *family* variable enables a descriptive reference to the whole malware.

1.2 Malware Detection Techniques

The evolution of malware detection techniques is analogous to an arms race against malware writers. When AV vendors deploy new methods to detect malware, the malware writers create new methods to usher their code past detection algorithms. Figure 2 provides a taxonomy of the standard malware detection techniques in use at present.

The main categories of analysis are outlined below:

1.2.1 Signature-Based

Signature-based detection relies on comparing the payload of a program against a repository of pre-learned regular expressions extracted from malware [14]. The extraction, storage and dissemination of these strings requires a large effort and a substantial amount of human time [13]. This detection is generally performed statically and so suffers from the major disadvantages of being ineffective against unknown regular expressions and being unable to search obfuscated code for such strings. This renders this technique increasingly impotent against an onslaught of exponentially growing malware.

1.2.2 Anomaly-Based

Anomaly-based detection systems classify what are considered typical behaviours for a file or system and any deviations from these are considered anomalies [41]. A training phase creates a model of typical behaviour or structure of the file or system. The monitoring phase then detects deviation from the baselines established in training [16]. For example, a poisoned PDF may differ greatly from the typical or expected structure of a clean PDF. Likewise, a one-hour TV show file may be expected to be around 350 MB in size. If the file was a mere 120 KB, it would be classed as suspicious. One of the major benefits of anomaly detection is the potential to detect zero-day threats. As the system is trained on typical traits, it does not need to know all of the atypical traits in advance of detection. Anomaly-based systems are, however, limited by propensity towards false-positive ratings and the scale of the features required to adequately model the system under inspection [16].

1.2.3 Specification-Based

Specification-based detection is a descendant of anomaly detection with a view to addressing the issue of high false-positive rates of the latter. This detection method seeks to create a rule set which approximates the requirements of the system, rather than the implementation of the system [41]. The adequate modelling of a large system is a difficult task and, as such, specification-based detection may have a similar disadvantage as to an anomaly-based system, in that the model does not adequately capture the behaviours of a complex system.

Each of these analysis types can be performed in one of three manners:

1.2.4 Static Analysis

Analysis is considered static when the subject is not executed and can be conducted on the source code or the compiled binary representation [13]. The analysis seeks to determine the function of the program, by using the code and data structures within [40]. MD5 hash representations of the executable can be checked against databases of hashes from previously detected malware. Call graphs can demonstrate the architecture of the software and the possible flows between functions. String analysis seeks out URL s, IP addresses, command line options, Windows Portable Executable files (PE) and passwords [29]. The main advantage of using static analysis techniques is that there is no threat from the malware sample, as it is never executed, thus can be safely analysed in detail. All code paths can potentially be examined, when the whole of the code is visible, whereas with dynamic analysis only the code being executed is examined. Examining all code paths is not always an advantage, as the information revealed may be redundant dead code, inserted to mask the true intentions of the file. Static analysis techniques are increasingly rendered redundant against growingly complex malware, which can obfuscate itself from analysis until the point of execution. As the code is never executed in static analysis, the obfuscation method does not always reveal the true code.

1.2.5 Dynamic Analysis

Dynamic analysis involves the execution of the file under investigation. Techniques mine information at the point of memory entry, during runtime and post-execution, therefore bypassing many of the obfuscation methods which stymie static analysis techniques. The dynamic environment can be: (a) *native*: where no separation between the host and the analysis environment is provided, as analysis takes place in the native host OS. With malware analysis, this obviously creates issues with the effects of the malware on the host. Software is available to provide a snapshot of the clean native environment, which can then be reloaded following each instance of malware execution, such as DeepFreeze. This can be quite a timely process however, as the entire system must be restored to the point of origin [23]; (b) *emulated*: where the host controls the guest environment through software emulating hardware. As this software is driven by the underlying host architecture, emulation can create performance issues; (c) *virtualised*: a virtual machine provides an isolated guest environment controlled by the host. In a perfect environment, the true nature of the subject is revealed and examinable. However, in retaliation, malware writers have created a battery of evasion techniques in order to detect an emulated, simulated or virtualized environment, including anti-debugging, anti-instrumentation and anti-Virtual Machine tactics. One major disadvantage of dynamic approaches is the runtime overhead in having to execute the program for a specified length of time. Static tools such as IDAPro can disassemble an executable in a few seconds, whereas dynamic analysis can take a few hundred times longer.

1.2.6 Hybrid Analysis

Hybrid techniques employ components of both static and dynamic analysis in their detection algorithms. In [34], Roundy and Miller used parsing to analyse suspect code statically pre-execution, building an analysis of the code structure, while dynamic execution was used for obfuscated code. The authors established an algorithm to intertwine dynamic and static techniques in order to instrument even obfuscated code.

1.3 Detection Evasion Techniques

Code obfuscation is the act of making code obscure or difficult to comprehend. Legitimate software houses have traditionally used obfuscation to try to prevent reverse engineering attempts on their products. This is the foundation of approaches such as Digital Rights Management, where emails, music etc are encrypted to prevent undesired use. Malware writers adopted these techniques to try to avoid detection of their code by AV scanners, in a similar fashion. In [11] the four most commonly used obfuscation techniques are listed as;

1. *Junk-insertion*: inserts additional code into a block, without altering its end behaviour. This can be dead code, such as *NOP* padding, irrelevant code, such as trivial mathematical functions and complex code with no purpose;
2. *Code transposition*: reorders the instructions in the code so that the resultant binary differs from the executable or from the predicted order of execution;
3. *Register reassignment*: substitutes the use of one register for another in a specific range;
4. *Instruction substitution*: takes advantage of the concept that there are many ways to do the same task in computer science. Using an alternative instruction to achieve the same end result provides an effective method for altering code appearance.

Packing software, or packers, are used for obfuscation, compression and encryption of a PE file, primarily to evade static detection methods. When loaded into RAM, the original executable is restored from its packed state in ROM, ready to unleash the payload on the victim machine [5].

Polymorphic (*many shapes*) malware is comprised of both the payload and an encryption/decryption engine. The polymorphic engine mutates the static code/-payload, which is decrypted at runtime back to the original code. This can cause problems for AV scanners, as the payload never appears the same in any descendant of the original code. However, the encryption/decryption engine potentially remains constant, and as such may be detected by signature- or pattern-matching scanners [26]. A mutation engine may be employed to obfuscate the decryption routine. This engine randomly generates a new decryption routine on each iteration, meaning the payload is encrypted and the decryptor is altered to appear different to the previously utilized decryptors [41].

Metamorphic (*changes shapes*) malware takes obfuscation a step further than polymorphism. With metamorphic malware, the entire code is rewritten on each iteration, meaning no two samples are exactly alike. As with polymorphic malware, the code is semantically the same, despite being different in construction or bytecode, with as few as six bytes of similarity between iterations [21]. Metamorphic malware can also use inert code excerpts from benignware to not only cause a totally different representation, but to make the file more like benignware, and thus evade detection [20].

1.4 Summary

The ongoing fight against malware seeks to find strategies to detect, prevent and mitigate malicious code before it can harm targeted systems. Specification-based and Anomaly-based detection methods model whole systems or networks for a baseline, and report changes (anomalous network traffic, registry changes etc), but are prone to high levels of false detections. Signature-detection is the most commonly used method among commercial AV applications [43]. With the ever-apparent evolution of malware, such approaches have been shown to be increasingly ineffective in the detection of malicious code [31]. This is particularly apparent in Zero Day exploits, where the time-lag between the malware's release and the generation of a signature leaves a significant window for the malcode to cause substantial damage. As these methods rely on a pre-learned signature, and as the ability of modern malware to obfuscate such signatures grows, signature-based AV approaches are by definition on a continuous back foot. Therefore, a new approach is required to be able to more accurately and efficiently detect malware which is unseen to the detection algorithm.

2 Related Research

Obfuscation has frustrated attempts at effective and efficient malware detection, both from a commercial application viewpoint and in a research context. Previous research has sought to find a successful malware detection algorithm by applying machine learning and data mining approaches to the problem. Schultz et al. [38] were the first researchers to present machine learning as a tool to investigate malicious binaries. Three classifiers (RIPPER, Naïve Bayes and Multi-Naïve Bayes) were implemented and compared to a signature-based algorithm, i.e. an AV scanner, using *program header*, *strings* and *byte sequences* as features. All of the machine learning approaches out-performed the AV scanner, with two of the classifiers doubling the accuracy of the commercial approach. Kolter and Maloof [20] successfully used a binary present/not present attribute for n-gram analysis of the hexadecimal representations of malware PE files. Among other classifiers assessed, Boosted Decision Trees gave an area under the receiver operating characteristic (AU-ROC) curve of 99.6%. In [27], features were used based on a host system to detect previously unseen worms. Data were collected on 323 features per second, focussing on the configuration, background activity and user activities of the host. Five worms were compared against a baseline behavioural set up, and with as few as 20 features, the average detection accuracy was >90%, with >99% for specific individual worms.

2.1 Opcode Analysis

Recent investigations into malware detection approaches have examined the behaviour of malware at run time, i.e. *what* the code is doing rather than *how* it is doing it, and how these behvaviours differ from benign code. *Opcodes* (**op**erational **codes**) are human-readable machine language instructions, issued directly to the CPU. Analysis at such a low-level provides the opportunity to detect malware, circumventing obfuscation at run time [31].

In [36], Santos et al. analysed n-gram combinations of opcodes in statically-generated datasets. An accuracy and an f-measure (the weighted average of precision over recall) of over 85% were observed using a ROC-SVM classifier, although n was limited to 2.

A graph-analysis approach was taken by [35] and applied to PE files and metamorphic malware, in an extension of [6]. Their model was able to detect metamorphic malware from benign applications, and metamorphic malware families from each other.

The authors in [31] analysed opcodes from a dynamically-yielded dataset using supervised machine learning. Run-traces of benign and malicious executables were taken by executing the applications using a virtual machine. This controlled environment allowed isolation of the guest OS, decoupling the effects of the malware from the host machine. A debugging program (Ollydbg), in conjunction with a masking tool (StrongOD), was used to capture the run-time trace of each program under investigation. This provided a raw trace file with each opcode and its operand, and further details such as a memory register address. A bespoke parser was used to extract the opcodes from this trace file and count their occurrences. The density of each opcode was calculated based on the frequency of the opcode within that sample. Different run-lengths were addressed by normalizing the opcode frequency using the total count within the sample. A pre-filter was used to select the features likely to provide the maximum information to the SVM, while minimizing the overall size problem created by use of n-gram analysis to investigate all combinations of opcodes. Principal components analysis (PCA) was used to rank each opcode by its importance to the SVM classification task. PCA compresses the data size while maintaining the variance in the data, allowing a subset of principal components to be generated. The researchers found that 99.5% of the variance in the data could be attributed to the top 8 opcodes, thus reducing the data from the original 150 opcodes, which was further validated by use of an SVM. However, the dataset used was only comprised of 300 benign and 350 malware PEs. Although previous research had fewer samples (e.g. [7]), low sample quantity and coverage have been recurring issues throughout the literature.

In [38], Schultz et al. used 3265 malware and 1001 benignware executables in the first study in this branch of malware literature. Santos et al. [37] considered 2000 files, evenly divided between classes. Both [28] and [39] used the same dataset, comprised of 7688 malicious files and 22,735 benign, listed by the researchers as the largest dataset in the literature at that time. However, there are notable limitations in the data collection methodology from these studies. The researchers stated that the disassembler used could only analyse 74% of their samples. Malware showed a greater attrition rate (26%) than benignware (10%), as the main reason for failure was the compression or packing of a file (i.e. obfuscation).

The dataset used by [36] contained 2000 files, due to unspecified technical limitations. The malware was randomly sampled from the 17,000-strong corpus of the VxHeaven website [2], though the static analysis used did not include packed files. The implication is that more sophisticated obfuscated malware was not used in the investigation, a considerable limitation of the study. Furthermore, >50% of the sample set was composed of three malware families: *hack-tool*, *back door* and *email worm*, which may have induced artificial within-class imbalances.

6721 malware samples were used in [17], harvested from VxHeaven and categorized into three types (*back door*, *worm*, and *trojan*), comprised of 26 malware families and >100 variations per family. No controls were used in the sampling across categories, with 497 worm, 3048 backdoor, and 3176 trojan variants, nor was there a methodological statement as to how these categories were sampled.

2.2   Rationale for the Current Research

Previous research in the field has typically used samples taken from the VxHeaven website, which are old and outdated, having last been updated in 2010. There are clear and notable limitations in the methodologies employed, with the majority having datasets which were not adequately sized, structured or sampled. Datasets which were generated statically failed to adequately address any serious form of obfuscation, with some just initially excluding any packed sample. While dynamic analysis does have disadvantages, it does offer a snapshot of the malware behaviour at runtime, regardless of obfuscation, and has been recommended by researchers who had previously focussed on static analysis [36]:

> Indeed, broadly used static detection methods can deal with packed malware only by using the signatures of the packers. As such, dynamic analysis seems like a more promising solution to this problem ([18]) [36, p.226]

3   Methodology

3.1   Source Data

Research in this field has typically used online repositories such as VxHeaven [2] or VirusShare [33]. The corpus of the former is dated, with the last updates being in 2010. VirusShare, however, contains many more samples, is continuously updated, and offers useful metadata. The site receives malware from researchers, security teams and the public, and makes it available for research purposes on an invite-only basis. The three most-recent folders of malware, at the time of writing, were downloaded, representing approximately 195,000 instances of malware, of varying formats. The Malicia dataset harvested from drive-by-downloads in the wild in [30] was also obtained and incorporated into the total malware corpus. Benign software was harvested from Windows machines, representing standard executables which would be encountered in a benchmark environment.

3.2 Database Creation

As the files in the malware corpus were listed by MD5 hash, with no file format information, a bespoke system was built to generate a database of attribute metadata for the samples. Consideration was given to using standard *nix commands such as *file* to yield the file formats. However, this did not provide all information necessary, and so a custom solution was engineered. Figure 3 illustrates the system employed.

A full-feature API key was granted by VirusTotal for this pre-processing stage. This allows a hash to be checked against the VirusTotal back-end database, which includes file definitions and the results of 54 AV scanners on that file. The generated report was parsed for the required features (Type, Format, First Seen, 54 AV scans etc) and written to a local database for storage. A final step in the attribution system allows the querying of the local database for specific file types and for files to be grouped accordingly. For the present research, we were interested in Windows PE files and our initial dataset yielded approximately 90,000 samples. Other file types (Android, PDF, JS etc) were categorized and stored for future research.

On creation of the local attribute dataset, there was a noticeable variation in whether each AV scanner determined that the specific file was malicious or not and also the designated malware type and family. This presented a challenge to our methodological stance. If a file is detected by very few of the 54 AV scanners, it could be viewed as a potential false positive (i.e. detected as malicious, though truly benign), or indeed that it is an instance of malware which can successfully evade being detected by most of the AV scanners. False positive detections can occur when a scanner mistakenly detects a file as malicious. This is exacerbated by the use of fuzzy hashing, where hash representations of a file are seen as similar to a malicious file, and the fact that scanning engines and signature databases are often shared by more than one vendor. The false positive issue can have serious consequences, such as rendering an OS unusable. It has become problematic enough that VirusTotal has attempted to build a whitelist of applications from their trusted sources, so that false positives can be detected and mitigated [24]. As such, we

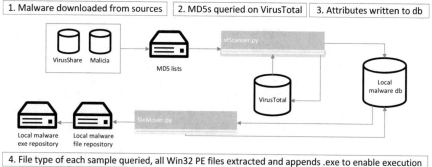

Fig. 3 Processing malware samples to store their descriptive attributes

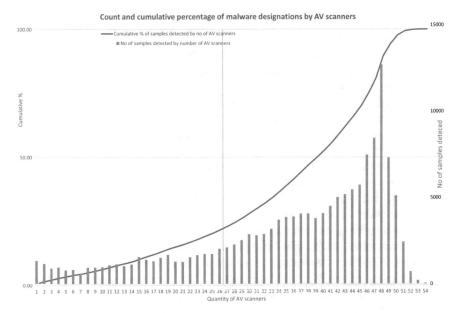

Fig. 4 The distribution of number of samples which were detected by the number of AV scanners (horizontal axis), with the cumulative percentage)

used a majority-rule algorithm and focussed on instances which were assessed as malware by 50% or more of the AV scanners. This can be justified by the fact that we wished to establish a fundamental malware dataset, with established malware being given priority over questionable samples. Indeed, the fact that there is disagreement between AV scanners over samples highlights one of the key issues in the field.

Figure 4 depicts the distribution of samples which had a specified number of AV scanners deem the sample as malicious. For example, 964 samples were judged to be malicious by 4 AV scanners from the VirusTotal report (right vertical axis), whereas 12,704 were detected by 48 scanners. Cumulatively, this represented 3% (left vertical axis) of the malware being detected by 4 or fewer scanners, and 89% by 48 or fewer. In Fig. 4, the central bisecting line represents the threshold after which malware was chosen for the experimental dataset, i.e. 26 or more of the AV scanners recognised the sample as malicious. This threshold demonstrates that 18% of the dataset was not included in the experimental dataset, by choosing to operate with a majority-rules decision. In other words, 82% of the malicious samples were retained while having the certainty of a rigorous inclusion mechanism. As the purpose of these experiments was to build a solid model of the dynamic behaviour of malware, the importance of widely agreed-upon samples for training purposes was a priority.

3.3 Automated Run-trace Collection

A process for extracting opcode run-traces from executed malware was established in [31, 32]. This generates the required traces, but is manually set up and operated, and so is unfeasible for generating a large dataset. As such, automated scalable processes were given consideration.

The work in [23] shows practical automated functions for use with VirtualBox virtual machines, though the methods used are now deprecated. A virtualization test-bed which resides outside of the host OS was developed by [12]. The authors claimed the tool was undetectable when tested by 25,118 malware samples, though no breakdown of the dataset is provided. However, the software was last updated in October 2015, is only compatible with Windows XP and requires a deprecated version of the Xen hypervisor, and so was not implemented.

The ubiquitous Cuckoo sandbox environment was considered also. While a valuable tool in malware and forensic analysis, we sought a system which provided the exact output for the method established in [31, 32]. Furthermore, we wish to develop an opcode-focussed malware analysis environment, and so chose to engineer a bespoke automated execution and tracing system.

VirtualBox was implemented as the virtualization platform. While other virtualization software exists, such as VirtualPC, Xen and VMWare, the API for VirtualBox is feature-rich and particularly suited to the present research. VirtualBox comprises multiple layers on top of the core hypervisor, which operates at the kernel level and permits autonomy for each VM from the host and other VMs. The API allows a full-feature implementation of the VirtualBox stack, both through the GUI and programmatically, and as such is well-documented and extensively tested [3].

A baseline image was created of Windows 10 64-bit, 2 GB RAM, VT-x acceleration, 2 cores of an Intel i5 5300 CPU, and the VBoxGuestAdditions v4.3.30 add-on installed. Windows 10 was selected as the latest guest OS due to the modernity and market share. VBoxGuestAdditions allows extra features to be enabled inside the VM, including folder-sharing and application execution from the host, both of which are key features of the implementation. Full internet access was granted to the guest and traffic was partially captured for separate research, and monitored as a further check for the liveness of the malware. Security features were minimised within the guest OS to maximise the effects of the executed malware.

OllyDbg v2 was preinstalled in the VM snapshot to trace the runtime behaviour of each executed file. This is an open-source assembler-level debugger, which can directly load both PE and DLL files and then debug them. The presence of a debugger can be detected by both malicious and benign files, and such instances may deploy masking techniques. As per [31], we used StrongOD v0.4.8.892 to cloak the presence of the debugger.

The guest OS was crafted to resemble a standard OS, with document and web histories, Java, Flash, .Net, non-empty recycling bin etc. While anti-anti-virtualization strategies were investigated, and implemented where appropriate, not all could be achieved while maintaining essential functionality, such as remote

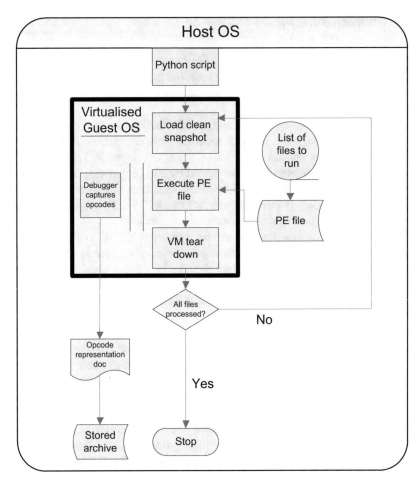

Fig. 5 Automated data collection system model

execution. However, while operating at the opcode level, we get the opportunity to monitor anti-virtualization behaviour and can turn this into a feature for detection. Furthermore, this research seeks to investigate the execution of malware as a user would experience it, which would include cloud-based virtualized machine instances. Figure 5 illustrates the workflow of the automated implementation used for the present research. On starting, a user-specified shared folder is parsed for all files which are listed for execution. The VM is spun-up using the baseline snapshot, mitigating any potential hang-overs from previously executed malware. OllyDbg is loaded with the masking plug-in and the next file from the execution list as a parameter. The debugger loads the PE into memory and pauses prior to execution. The debugger is set to live overwrite an existing empty file with the run-trace of opcode instructions which have been sent to the CPU, set at a trace-into level so that every step is captured. This is in contrast to static analysis, which only yields

Address	Thread	Command	; Registers an(
00450E11	Main	MOV ESI,Trojan_W.00433000	; ESI=00433000
00450E16	Main	LEA EDI,DWORD PTR DS:[ESI+FFFCE000]	; EDI=00401000
00450E1C	Main	MOV DWORD PTR DS:[EDI+404CC],54200703	
00450E26	Main	PUSH EDI	
00450E27	Main	OR EBP,FFFFFFFF	; EBP=FFFFFFFF
00450E2A	Main	JMP SHORT Trojan_W.00450E3A	
00450E3A	Main	MOV EBX,DWORD PTR DS:[ESI]	; EBX=FFFE665B

Fig. 6 Sample lines from a runtime trace file

potentially executed code and code-paths. Figure 6 shows an example run-trace file which has been yielded. After the set run-length, which was set to 9 min, elapsed, the VM is torn down and the process restarts with the next file to be executed until the list is exhausted. A 9 min execution time was selected, with 1 min for start up and teardown, as we are focussed on detecting malware as accurately as possible, but within as short a runtime as possible. While a longer runtime may potentially yield richer trace files and more effectively trace malware with obfuscating sleep functions, our research concentrates on the initial execution stage for detection processes.

The bottleneck in dynamic execution is run-time, so any increase in capture rate will be determined by parallelism while run-time is held constant. Two key benefits of the implementation presented are the scalability and automated execution. The initial execution phase was distributed across 14 physical machines, allowing the dataset to be processed in parallel. This allowed the creation of such a large dataset, in the context of the literature, within a practical period of time. Indeed, this was only capped at the number of physical machines available. It is entirely feasible to have a cloud- or server- based implementation of this tracing system, with a virtually unlimited number of nodes.

3.4   Opcode Statistics Collection

Prior to analysis, the corpus of trace files require parsing to extract the opcodes in sequence. A bespoke parser was employed to scan through each run-trace file and keep a running count of the occurrence of each of the 610 opcodes listed in the Intel x86/x64 architecture [22]. This list is more comprehensive than that of [31, 32], as previous research has indicated that more rarely occurring opcodes can provide better discrimination between malware and benignware than more frequently occurring opcodes [7, 17].

As the run-traces vary in length (i.e. number of lines), a consistent run-length is needed for the present research in order to investigate the effects of run-length on malware detection. A file slicer was used on the run-trace dataset to truncate all files to the maximum required in order to control run-length. As partly per [31, 32],

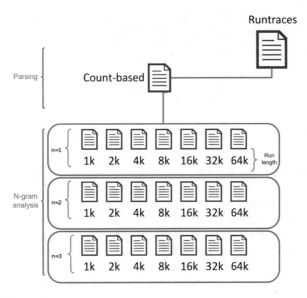

Fig. 7 Derived datasets

	n = 1		n = 2		n = 3
MOV	} n-gram 1	MOV	} n-gram 1	MOV	
PUSH	} n-gram 2	PUSH	}	PUSH	} n-gram 1
POP	} n-gram 3	POP	} n-gram 2	POP	

Fig. 8 Sample n-gram combinations

run-lengths were capped at 1, 2, 4, 8, 16, 32 and 64 thousand opcodes in sequence, along with the non-truncated set enumerated in Fig. 7.

The count for each trace file is appended to the overall dataset and stored as a CSV file for analysis.

As the behaviour of the malware at runtime is represented textually, it can be seen as a natural-language problem, with a defined dictionary. N-grams are used in text processing for machine learning, which can equally be applied to the problem at hand. Previous research has shown that n-gram analysis can be used to discriminate malicious and benign applications, though datasets have been severely limited. As such, this paper focuses on n-gram analysis to investigate classification of malware on a substantial dataset.

A second separate phase of the opcode counting process then parses the run-traces for bi-grams and trigrams from the reduced list ($n = 2..3$). The system removes all opcodes which did not occur, as with the increase in n, the possible combinations rise exponentially. As the opcodes which did not occur at $n = 1$ cannot be in combinations of $n = 2$ etc, this can mitigate the feature explosion without losing any data. This is depicted in Fig. 8.

Table 1 Composition of dataset

Malware category	Samples
Adware	592
Backdoor	1031
BrowserModifier	721
Malicia	8242
Null	5491
PWS	1964
Ransom	76
Rogue	498
SoftwareBundler	630
Trojan	6018
TrojanDownloader	3825
TrojanDropper	1619
TrojanSpy	405
VirTool	2206
Virus	6341
Worm	7251
Benign	1065
	47,975

From initial download through to execution, the number of samples was depleted by approximately 20%. A further 10 % of files failed to provide a run-trace for a variety of reasons, such as missing dlls or lock-outs (Table 1).

The quantity of samples and features in any large dataset can provide methodological challenges for analyses, which was pertinent in the present study. For example, with the $n = 3$ *64k run-length* dataset, there were 47,975 instances, with 100,318 features, or 4.82 *billion* data points. When contained within a CSV format, the file size approaches 9.4 GB. All analyses were conducted on a server-class machine with a 12-core Intel Xeon CPU and 96 GB of RAM. Despite this computational power, datasets of such magnitude are problematic, and so a sparse-representation file format was chosen (i.e. zeros were explicitly removed from the dataset, implied by their omission). This reduced the dataset file sizes by as much as 90%, indicating a large quantity of zero entries, and so feature reduction was investigated, as discussed further below.

3.5 Application of Machine Learning

All malware types were merged into a single malware category and compared to the benign class. Due to the size imbalance, a hybrid over/under/sub- sampling approach was taken to balance the classes. The minority (benign) class was synthetically oversampled with the Synthetic Minority Oversampling Technique (SMOTE) [10]. This approach uses a reverse-KNN algorithm to generate new instances with values

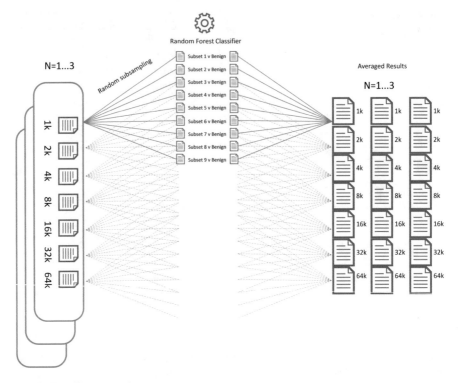

Fig. 9 Sampling approach

based on the values of 5 nearest neighbours, generating a 400% increase in benign sample size in our dataset. The majority class was then sub-sampled randomly into ninefolds, with a matching algorithm ensuring datasets were of equal size. Figure 9 illustrates this approach. This ensures that the training model was built with data which were balanced between the two classes, to try to mitigate negative impacts on model performance which imbalanced data can have.

The Random Forest (RF) [9] classification algorithm was used, as implemented in WEKA 3.9, to classify the data in all experiments. RF is an ensemble learner, combining the decisions of multiple small learners (decision trees). RF differs from traditional tree-based learning algorithms, as a random number of features are employed at each node in the tree to choose the parameter, which increases noise immunity and reduces over-fitting. RF functions well with datasets containing large numbers of instances and features, particularly with wide datasets (i.e. more features than instances). It parallelizes well [9], and is also highly recommended with imbalanced data [19]. In pilot experiments, RF provided high accuracy, despite class imbalances and short run-lengths, even with the large number of features being used. Tenfolds cross-validation was employed on the classification of each subset, and the average of the results over the nine subsets was taken to produce the overall results. Parameter selection for *number of trees* and *number of features* was conducted initially by a coarse grid search, followed by a fine grid search. For both parameters higher is better, especially over large datasets, though classification will plateau and a greater computational overhead penalty will be enacted.

4 Results

Standard machine learning outputs were yielded in the evaluation of the classifier for each dataset:

1. Accuracy, i.e. the percentage of correctly assigned classes (0% indicating a poor classifier, 100% showing perfect accuracy)

$$ACC = \frac{TP + TN}{TotalPopulation}$$

2. F_1 score is the harmonic mean of precision and sensitivity (1 is the ideal).

$$F_1 = \frac{2TP}{2TP + FP + FN}$$

3. Area Under Receiver Operating Characteristic curve, i.e. when the True Positive % rate is plotted against the False Positive % rate, the area underneath this curve (1 is the ideal).
4. Area Under Precision Recall curve (1 is the ideal) where

$$Precision = \frac{TP}{TP + FP}$$
$$Recall = \frac{TP}{TP + FN}$$

4.1 Effects of Run-Length and n-gram Size on Detection Rates

The evaluations of each classification, prior to any feature reduction, are depicted in Fig. 10.

The overall accuracy across all 21 analyses had a mean of 95.87%, with a range of 93.01%–99.05%. The accuracy level declined in 3 of the 7 categories, between $n = 2$ and $n = 3$, though increased in 3 and remained stable in one. This indicates that there is marginal gain (0.06% on average) in increasing n from 2 to 3. The $n = 1$ cohort did, however, show an overall higher level of accuracy in all but one run-length (16k) and a greater mean by 1.29% vs $n = 2$ and 1.23% vs $n = 3$. F-scores ranged from 0.93 to 0.991, with the 3 n-gram means again showing greater

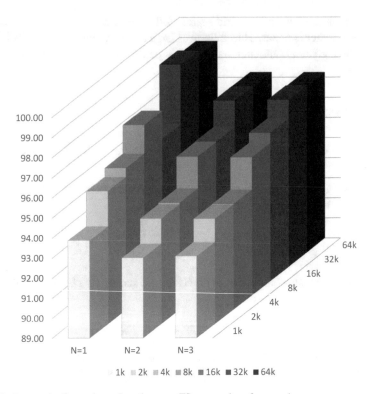

Fig. 10 Accuracies for each run-length across FS approaches, for $n = 1$

performance by $n = 1$, with marginal difference between $n = 2$ and $n = 3$. The AU-ROC showed high values, >0.979, with $n = 1$ from 0.99 to 0.999.

All measures increased as run-length increased, up to 32k. At 64k, all measures were either static or showed a marginal decline in performance, though most notably in accuracy (from 99.05% to 98.82%). Across all measures, the overall best performing strand was $n = 1$ at 32k run-length with an accuracy of 99.05%, f-score of 0.991 and AU-ROC of 0.999.

4.2   Feature Selection and Extraction

With the large quantity of features, attempts were made to reduce the number of attributes, while maintaining the detection capability of the model. Furthermore, it would not be computationally possible to employ feature extraction algorithms within a feasible time period, with such a large feature set. As such, a feature selection strategy was employed in the first instance.

Table 2 Quantity of features remaining per run-length for each FS category

	No FS			GR ≥0.01			Top 20			Top 50		
	$n = 1$	$n = 2$	$n = 3$	$n = 1$	$n = 2$	$n = 3$	$n = 1$	$n = 2$	$n = 3$	$n = 1$	$n = 2$	$n = 3$
1k	610	1023	66,443	111	912	10,103	20	20	20	50	50	50
2k	610	6075	70,777	111	2448	12,270	20	20	20	50	50	50
4k	610	6358	75,907	115	2597	14,220	20	20	20	50	50	50
8k	610	6677	80,876	116	2734	15,804	20	20	20	50	50	50
16k	610	6956	85,693	135	2909	17,493	20	20	20	50	50	50
32k	610	7215	91,403	119	2947	19,533	20	20	20	50	50	50
64k	610	8135	100,316	139	3532	22,796	20	20	20	50	50	50

The Gain Ratio algorithm was employed to reduce the number of features in the datasets. Gain Ratio attempts to address the biases which can be introduced by Information Gain, when considering attributes with a large range of distinct values, by considering the number of branches that would result due to a split. In the ratio, the numerator represents the information yielded about the label and the denominator represents the information learned about the feature. The *ranker* search method was used to traverse the feature set. With respect to the current problem, *ranker* is advantageous as it is linear and does not take more steps than the number of features. With the breadth of the present datasets, this provides a saving of computational overhead.

Table 2 shows the quantity of features in each cohort and run-length at each step. The number of features rises in $n = 2$ and $n = 3$, and in each increase in run-length, as the potential number of observed opcodes rises.

Four levels of feature selection using Gain Ratio were investigated: no feature selection, Gain Ratio merit score >=0.01 (i.e. non-zero), top 20 features, and top 50 features. A further category was originally investigated (merit score >=10) however, this resulted in a model equivalent to random choice (i.e. all measures scored 50%). The same RF classifier was then presented with the reduced datasets for classification.

Removing features with a merit score of <0.01 had a negligible impact on the accuracy rating (0.013–0.018%) compared to the full feature set, but removed up to 84% of features.

Selecting the top n features was subsequently investigated, as the removal of non-zero features is unique to each dataset but *top n* would provide uniformity. The $n = 1$ dataset proved more resilient across top 20 and top 50 features than the $n = 2$ and $n = 3$ cohorts. This may be due to the difference in quantity of features from no reduction to top n, e.g. $n = 3$ 64k was reduced from 100,316 to 20 features in this approach. Figure 11 shows accuracy for $n = 1$.

With any malware detection model using machine learning, consideration should be given to the false positive rates (FP) (i.e. detected incorrectly as a specific class). This is particularly relevant with imbalanced data. Table 3 lists the false positive ratings for both the benign and malicious classes, per run-length and n-gram size.

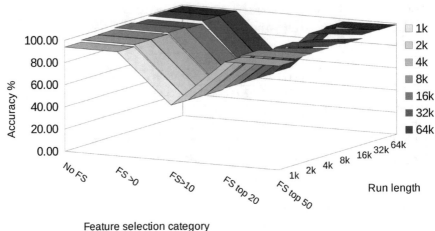

Fig. 11 Accuracies for each run-length across FS approaches, for $n = 1$

Table 3 False positive rates for each run-length for $n = 1..3$ using all features

	$n = 1$			$n = 2$			$n = 3$		
	Benign	Malicious	Mean	Benign	Malicious	Mean	Benign	Malicious	Mean
1k	0.042	0.081	0.061	0.029	0.111	0.070	0.031	0.107	0.069
2k	0.040	0.048	0.044	0.036	0.079	0.057	0.040	0.075	0.058
4k	0.046	0.032	0.039	0.046	0.068	0.057	0.044	0.071	0.057
8k	0.022	0.028	0.025	0.023	0.059	0.041	0.024	0.059	0.041
16k	0.215	0.027	0.121	0.011	0.044	0.027	0.020	0.052	0.036
32k	0.008	0.011	0.009	0.008	0.047	0.027	0.010	0.045	0.027
64k	0.013	0.011	0.012	0.022	0.060	0.041	0.011	0.044	0.028
Mean	0.055	0.034	0.044	0.025	0.067	0.046	0.026	0.065	0.045

The 'Benign' columns quantify the rate of the model labeling the sample incorrectly as benign, and vice versa for the 'Malicious' columns. Overall FP rates are low, though a range of values is evident, consistent with the other machine learning measures. Again, the $n = 1$ data performs slightly better than the other n-gram sizes (mean=0.044 vs means of 0.046 and 0.045 respectively). The 32k dataset shows the lowest false positive rates (benign 0.008 and malicious 0.011, giving a mean of <1%). This indicates the strength of the model for accurate detection, while controlling for false positives. Furthermore, the sampling approach taken to balance the training data multiple times appears to have reduced the risk of increased false positives with unbalanced datasets (Fig. 12).

The further metrics for the model are presented in Table 4. As run-length increases, model performance increases for the *No FS* and *merit>0* cohorts, with the exception of the anomalous 16k set. The $n = 2$ and $n = 3$ datasets decline in performance as run-length increases when a set number of features are used (*Top 20*

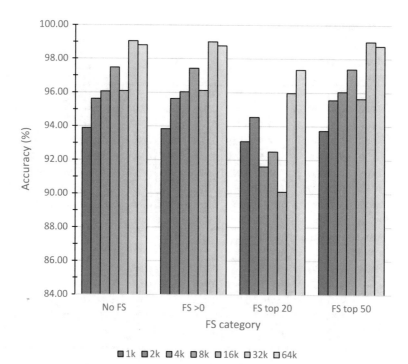

■1k ■2k ■4k ■8k □16k □32k □64k

Fig. 12 Mean accuracies for each run-length across FS approaches for all n-gram lengths

and *Top 50*). In particular the AU-ROC scores show a decline as run-length increases for $n = 2$ using 20 features. An intuitive explanation would be due to the ratio of original to selected feature numbers. However, of the three n-gram sizes, $n = 2$ had the lowest average percentage loss of features after the GR>0 feature selection ($n = 1$: 80.19%, $n = 2$: 51.81%, $n = 3$: 80.67%). It had approximately 10 times fewer features than $n = 3$ in the original cohort, and so the reduction to a fixed 20 or 50 should have had more of an impact if feature ratio was the important factor. As AU-ROC measures discrimination, it would appear that the information for correctly discriminating between benign and malicious code is distributed differently when bi-grams are inspected.

5 Conclusions

With accuracy of 99.05% across such a large dataset, and using tenfold cross-validation, there is a clear indication that dynamic opcode analysis can be used to detect malicious software in a practical application. While the time taken by dynamic analysis can be a disadvantage, the work presented here shows that a trace with as few as 1000 opcodes can correctly discriminate malware from benign with

Table 4 Model metrics per run-length, n-gram length and FS method

No feature selection

	Accuracy (% correct)			F-score			ROC			PRC		
	$n = 1$	$n = 2$	$n = 3$	$n = 1$	$n = 2$	$n = 3$	$n = 1$	$n = 2$	$n = 3$	$n = 1$	$n = 2$	$n = 3$
1k	93.88	93.01	93.09	0.939	0.93	0.931	0.99	0.983	0.979	0.99	0.983	0.979
2k	95.62	94.26	94.24	0.956	0.943	0.942	0.994	0.986	0.984	0.994	0.986	0.983
4k	96.07	94.32	94.26	0.961	0.943	0.943	0.994	0.987	0.984	0.993	0.986	0.984
8k	97.48	95.91	95.87	0.975	0.959	0.964	0.996	0.99	0.988	0.995	0.99	0.986
16k	96.09	95.94	96.37	0.961	0.959	0.964	0.991	0.991	0.988	0.99	0.991	0.986
32k	99.05	97.26	97.30	0.991	0.973	0.973	0.999	0.994	0.991	0.998	0.993	0.99
64k	98.82	97.25	97.25	0.988	0.973	0.972	0.998	0.993	0.99	0.998	0.993	0.989
Mean	96.72	95.42	95.48	0.97	0.95	0.96	0.99	0.99	0.99	0.99	0.99	0.99

After Feature Selection using GainRatio with all features showing merit >0

	Accuracy (% correct)			F-score			ROC			PRC		
	$n = 1$	$n = 2$	$n = 3$	$n = 1$	$n = 2$	$n = 3$	$n = 1$	$n = 2$	$n = 3$	$n = 1$	$n = 2$	$n = 3$
1k	93.84	92.97	93.09	0.938	0.93	0.931	0.99	0.983	0.979	0.99	0.982	0.979
2k	95.64	94.26	94.24	0.956	0.943	0.942	0.994	0.986	0.984	0.994	0.986	0.983
4k	96.04	94.27	94.24	0.96	0.943	0.942	0.994	0.987	0.984	0.993	0.986	0.983
8k	97.44	95.88	95.81	0.974	0.959	0.958	0.996	0.99	0.987	0.995	0.99	0.986
16k	96.13	95.95	96.32	0.961	0.96	0.963	0.991	0.991	0.988	0.991	0.991	0.987
32k	99.01	97.29	97.30	0.99	0.973	0.973	0.999	0.993	0.991	0.999	0.993	0.99
64k	98.78	97.23	97.26	0.988	0.972	0.973	0.998	0.993	0.99	0.998	0.993	0.99
Mean	96.70	95.41	95.47	0.97	0.95	0.95	0.99	0.99	0.99	0.99	0.99	0.99

After Feature Selection using GainRatio with top 20 features

	Accuracy (% correct)			F-score			ROC			PRC		
	$n = 1$	$n = 2$	$n = 3$	$n = 1$	$n = 2$	$n = 3$	$n = 1$	$n = 2$	$n = 3$	$n = 1$	$n = 2$	$n = 3$
1k	93.10	82.59	83.95	0.931	0.821	0.837	0.987	0.841	0.859	0.986	0.827	0.845
2k	94.54	83.27	82.28	0.945	0.829	0.818	0.989	0.843	0.828	0.988	0.832	0.815
4k	91.61	83.72	83.21	0.916	0.833	0.828	0.971	0.844	0.845	0.97	0.833	0.833
8k	92.50	81.54	81.90	0.925	0.809	0.813	0.972	0.814	0.82	0.971	0.797	0.807
16k	90.11	81.60	82.28	0.901	0.811	0.817	0.953	0.816	0.824	0.949	0.798	0.812
32k	95.97	62.92	82.73	0.96	0.57	0.822	0.987	0.627	0.827	0.986	0.629	0.812
64k	97.35	62.84	82.33	0.974	0.569	0.818	0.99	0.624	0.823	0.989	0.627	0.807
Mean	93.60	76.93	82.67	0.94	0.75	0.82	0.98	0.77	0.83	0.98	0.76	0.82

After Feature Selection using GainRatio with top 50 features

	Accuracy (% correct)			F-score			ROC			PRC		
	$n = 1$	$n = 2$	$n = 3$	$n = 1$	$n = 2$	$n = 3$	$n = 1$	$n = 2$	$n = 3$	$n = 1$	$n = 2$	$n = 3$
1k	93.74	85.23	84.57	0.937	0.85	0.844	0.99	0.889	0.872	0.989	0.885	0.862
2k	95.57	86.03	82.80	0.956	0.858	0.824	0.994	0.884	0.837	0.993	0.878	0.825
4k	96.05	84.93	83.96	0.96	0.847	0.836	0.993	0.882	0.86	0.992	0.879	0.849
8k	97.39	84.65	82.47	0.974	0.844	0.82	0.995	0.869	0.827	0.995	0.862	0.815
16k	95.64	84.10	82.73	0.956	0.838	0.822	0.987	0.869	0.829	0.987	0.861	0.817
32k	99.01	85.98	82.79	0.99	0.857	0.823	0.999	0.89	0.828	0.998	0.881	0.813
64k	98.75	79.87	82.65	0.988	0.791	0.822	0.998	0.805	0.826	0.998	0.796	0.809
Mean	96.59	84.40	83.14	0.97	0.84	0.83	0.99	0.87	0.84	0.99	0.86	0.83

93.8% success. Increasing the run-trace to the first 32,000 opcodes increases the accuracy to over 99%. In terms of execution time, this run-length would not be dissimilar to static tools such as IDA Pro. Further reducing the feature set to the top 50 attributes and using a unigram representation upholds the classification accuracy >99%, while reducing the parsing, processing, and classification overheads. In terms of n-gram analysis, there is minimal gain when increasing between $n = 2$ and $n = 3$. Furthermore, n>1 offers no increase in detection accuracy when runtime is controlled for. Considering the computational overhead and feature explosion with increasing levels of n, focus should be maintained on a unigram investigation.

The results presented concur with [32], who found 32k run-length to provide the highest accuracy. However, the accuracy of the model peaked at 86.31%, with 13 features being included. Furthermore, the present study employs a dataset approximately 80 times the size of [32]. In the context of previous similar research, the accuracy of the present model is superior while the dataset is significantly larger and more representative of malware.

Acknowledgements The authors wish to acknowledge the extra support of VirusTotal.com by providing enhanced capabilities on their system.

References

1. 2014 cost of data breach study. Tech. rep., Ponemon Inst, IBM (2014). URL http://public.dhe. ibm.com/common/ssi/ecm/se/en/sel03027usen/SEL03027USEN.PDF
2. Vxheaven (2014). URL http://vxheaven.org/vl.php
3. Oracle vm virtualbox r programming guide and reference. Tech. rep., Oracle Corp. (2015b). URL http://download.virtualbox.org/virtualbox/SDKRef.pdf
4. Mcafee labs threats report sept 2016. Tech. rep., McAfee Labs (2016). URL http://www. mcafee.com/us/resources/reports/rp-quarterly-threat-q1-2015.pdf
5. Alazab, M., Venkatraman, S., Watters, P., Alazab, M., Alazab, A.: Cybercrime: The case of obfuscated malware, pp. 204–211. Global Security, Safety and Sustainability & e-Democracy. Springer (2012)
6. Anderson, B., Quist, D., Neil, J., Storlie, C., Lane, T.: Graph-based malware detection using dynamic analysis. Journal in Computer Virology 7(4), 247–258 (2011)
7. Bilar, D.: Opcodes as predictor for malware. International Journal of Electronic Security and Digital Forensics 1(2), 156–168 (2007)
8. Bontchev V., S.F., Solomon, A.: Caro virus naming convention (1991)
9. Breiman, L.: Random forests. Machine Learning 45(1), 5–32 (2001)
10. Chawla, N., Japkowicz, N., Kolcz, A.: Special issue on class imbalances. SIGKDD Explorations 6(1), 1–6 (2004)
11. Christodorescu, M., Jha, S.: Static analysis of executables to detect malicious patterns. In: Proceedings of the 12th Usenix Security Symposium, pp. 169–185. USENIX Association (2006)
12. Dinaburg, A., Royal, P., Sharif, M., Lee, W.: Ether: malware analysis via hardware virtualization extensions. In: Proceedings of the 15th ACM conference on Computer and communications security, pp. 51–62. ACM (2008)
13. Egele, M., Scholte, T., Kirda, E., Kruegel, C.: A survey on automated dynamic malware-analysis techniques and tools. ACM Computing Surveys 44(2), 6–6:42 (2012). DOI

10.1145/2089125.2089126.   URL http://search.ebscohost.com/login.aspx?direct=true&db=
buh&AN=77698357&site=eds-live&scope=site

14. Ellis, D.R., Aiken, J.G., Attwood, K.S., Tenaglia, S.D.: A behavioral approach to worm
    detection. In: Proceedings of the 2004 ACM workshop on Rapid malcode, pp. 43–53. ACM
    (2004)
15. FitzGerald, N.: A virus by any other name: Towards the revised caro naming convention.
    Proc.AVAR pp. 141–166 (2002)
16. Idika, N., Mathur, A.P.: A survey of malware detection techniques. Purdue University **48**
    (2007)
17. Kang, B., Han, K.S., Kang, B., Im, E.G.: Malware categorization using dynamic mnemonic
    frequency analysis with redundancy filtering. Digital Investigation **11**(4), 323–335 (2014).
    DOI http://dx.doi.org.queens.ezp1.qub.ac.uk/10.1016/j.diin.2014.06.003
18. Kang, M.G., Poosankam, P., Yin, H.: Renovo: A hidden code extractor for packed executables.
    In: Proceedings of the 2007 ACM workshop on Recurring malcode, pp. 46–53. ACM (2007)
19. Khoshgoftaar, T.M., Golawala, M., Hulse, J.V.: An empirical study of learning from imbal-
    anced data using random forest. In: 19th IEEE International Conference on Tools with Artificial
    Intelligence (ICTAI 2007), vol. 2, pp. 310–317. IEEE (2007)
20. Kolter, J.Z., Maloof, M.A.: Learning to detect malicious executables in the wild.   In:
    Proceedings of the tenth ACM SIGKDD international conference on Knowledge discovery
    and data mining, pp. 470–478. ACM (2004)
21. Leder, F., Steinbock, B., Martini, P.: Classification and detection of metamorphic malware
    using value set analysis. In: Malicious and Unwanted Software (MALWARE), 2009 4th
    International Conference on, pp. 39–46. IEEE (2009)
22. Lejska, K.: X86 opcode and instruction reference. URL http://ref.x86asm.net/
23. Ligh, M., Adair, S., Hartstein, B., Richard, M.: Malware analyst's cookbook and DVD: tools
    and techniques for fighting malicious code. Wiley Publishing (2010)
24. Martinez, E.: A first shot at false positives (2015). http://blog.virustotal.com/2015/02/a-first-
    shot-at-false-positives.html
25. McGraw, G., Morrisett, G.: Attacking malicious code: A report to the infosec research council.
    IEEE Software (5), 33–41 (2000)
26. Mehra, V., Jain, V., Uppal, D.: Dacomm: Detection and classification of metamorphic malware.
    In: Communication Systems and Network Technologies (CSNT), 2015 Fifth International
    Conference on, pp. 668–673. IEEE (2015)
27. Moskovitch, R., Elovici, Y., Rokach, L.: Detection of unknown computer worms based
    on behavioral classification of the host. Computational Statistics & Data Analysis **52**(9),
    4544–4566 (2008)
28. Moskovitch, R., Feher, C., Tzachar, N., Berger, E., Gitelman, M., Dolev, S., Elovici, Y.:
    Unknown malcode detection using OPCODE representation, pp. 204–215. Intelligence and
    Security Informatics. Springer (2008)
29. Namanya, A.P., Pagna-Disso, J., Awan, I.: Evaluation of automated static analysis tools
    for malware detection in portable executable files. In: 31st UK Performance Engineering
    Workshop 17 September 2015, p. 81 (2015)
30. Nappa, A., Rafique, M.Z., Caballero, J.: Driving in the cloud: An analysis of drive-by
    download operations and abuse reporting, pp. 1–20. Detection of Intrusions and Malware,
    and Vulnerability Assessment. Springer (2013)
31. O'kane, P., Sezer, S., McLaughlin, K., Im, E.G.: Svm training phase reduction using dataset
    feature filtering for malware detection. IEEE transactions on information forensics and security
    8(3-4), 500–509 (2013)
32. O'kane, P., Sezer, S., McLaughlin, K., Im, E.G.: Malware detection: program run length against
    detection rate. IET software **8**(1), 42–51 (2014)
33. Roberts, J.M.: VirusShare.com (2014)
34. Roundy, K.A., Miller, B.P.: Hybrid analysis and control of malware. In: Recent Advances in
    Intrusion Detection, pp. 317–338. Springer (2010)

35. Runwal, N., Low, R.M., Stamp, M.: Opcode graph similarity and metamorphic detection. Journal in Computer Virology **8**(1-2), 37–52 (2012)
36. Santos, I., Brezo, F., Sanz, B., Laorden, C., Bringas, P.G.: Using opcode sequences in single-class learning to detect unknown malware. IET information security **5**(4), 220–227 (2011)
37. Santos, I., Sanz, B., Laorden, C., Brezo, F., Bringas, P.G.: Opcode-sequence-based semi-supervised unknown malware detection, pp. 50–57. Computational Intelligence in Security for Information Systems. Springer (2011)
38. Schultz, M.G., Eskin, E., Zadok, E., Stolfo, S.J.: Data mining methods for detection of new malicious executables. In: Security and Privacy, 2001. S&P 2001. Proceedings. 2001 IEEE Symposium on, pp. 38–49. IEEE (2001)
39. Shabtai, A., Moskovitch, R., Feher, C., Dolev, S., Elovici, Y.: Detecting unknown malicious code by applying classification techniques on opcode patterns. Security Informatics **1**(1), 1–22 (2012)
40. Sikorski, M., Honig, A.: Practical Malware Analysis: The Hands-On Guide to Dissecting Malicious Software. No Starch Press (2012)
41. Thunga, S.P., Neelisetti, R.K.: Identifying metamorphic virus using n-grams and hidden markov model. In: Advances in Computing, Communications and Informatics (ICACCI), 2015 International Conference on, pp. 2016–2022. IEEE (2015)
42. Veluz, D.: Stuxnet malware targets scada systems (2010). https://www.trendmicro.com/vinfo/us/threat-encyclopedia/web-attack/54/stuxnet-malware-targets-scada-systems
43. Vemparala, S., Troia, F.D., Corrado, V.A., Austin, T.H., Stamp, M.: Malware detection using dynamic birthmarks. In: IWSPA 2016 - Proceedings of the 2016 ACM International Workshop on Security and Privacy Analytics, co-located with CODASPY 2016, pp. 41–46 (2016). DOI 10.1145/2875475.2875476

Big Data Analytics for Intrusion Detection System: Statistical Decision-Making Using Finite Dirichlet Mixture Models

Nour Moustafa, Gideon Creech, and Jill Slay

Abstract An intrusion detection system has become a vital mechanism to detect a wide variety of malicious activities in the cyber domain. However, this system still faces an important limitation when it comes to detecting zero-day attacks, concerning the reduction of relatively high false alarm rates. It is thus necessary to no longer consider the tasks of monitoring and analysing network data in isolation, but instead optimise their integration with decision-making methods for identifying anomalous events. This chapter presents a scalable framework for building an effective and lightweight anomaly detection system. This framework includes three modules of capturing and logging, pre-processing and a new statistical decision engine, called the Dirichlet mixture model based anomaly detection technique. The first module sniffs and collects network data while the second module analyses and filters these data to improve the performance of the decision engine. Finally, the decision engine is designed based on the Dirichlet mixture model with a lower-upper interquartile range as decision engine. The performance of this framework is evaluated on two well-known datasets, the NSL-KDD and UNSW-NB15. The empirical results showed that the statistical analysis of network data helps in choosing the best model which correctly fits the network data. Additionally, the Dirichlet mixture model based anomaly detection technique provides a higher detection rate and lower false alarm rate than other three compelling techniques. These techniques were built based on correlation and distance measures that cannot detect modern attacks which mimic normal activities, whereas the proposed technique was established using the Dirichlet mixture model and precise boundaries of interquartile range for finding small differences between legitimate and attack vectors, efficiently identifying these attacks.

N. Moustafa (✉) • G. Creech • J. Slay
The Australian Centre for Cyber Security, University of New South Wales Canberra, Canberra, NSW, Australia
e-mail: nour.moustafa@unsw.edu.au; G.Creech@adfa.edu.au; j.slay@adfa.edu.au

© Springer International Publishing AG 2017
I. Palomares et al. (eds.), *Data Analytics and Decision Support for Cybersecurity*,
Data Analytics, DOI 10.1007/978-3-319-59439-2_5

127

1   Introduction

In the cyber security field, an Intrusion Detection System (IDS) is essential for achieving a solid line of defence against cyber intrusions. The digital world has become the principal complement to the physical world because of the widespread usage of computer networks and prevalence of programs and services which easily accomplish users' tasks in a short time at a low cost. A system is considered secure if the three principles of computer security, Confidentiality, Integrity and Availability (CIA), are successfully satisfied [38, 43]. Hackers always endeavour to violate these principles, with each attack type having its own sophisticated manner and posing serious threats to computer networking.

An Anomaly-based Detection System (ADS), a specific methodology of IDS discussed in Sect. 2, still faces two problems for implementation in large-scale industrial applications [2, 7, 39], such as cloud computing [30] and SCADA systems [16]. Firstly, and most importantly, the construction of a profile from various legitimate patterns is extremely difficult because of the frequent changes of the normal data [2, 34, 47]. Secondly, the process of building a scalable, adaptive and lightweight detection method is an arduous task with the high speeds and large sizes of current networks [39, 47].

ADS methodologies have been developed using approaches involving data mining and machine learning, artificial intelligence, knowledge-based and statistical models [7, 34, 39]. Nevertheless, usually, these proposed techniques have produced high False Positive Rates (FPRs) because of the difficulty of designing a solution which solves the above problems. Recent research studies [27, 34, 41, 44, 47] have focused on statistical models due to the ease of concurrently employing and determining the potential properties of both normal and abnormal patterns of behaviour. Discovering these properties and characterising a certain threshold for any detection method to correctly detect attacks requires accurate analysis.

Both network and host systems have multiple devices, software, sensors, platforms and other sources connected together to deliver services to users and organisations anytime and anywhere. In addition, these systems monitor the demands from such organisations by using Big Data analytical techniques and tools to carefully provide decision support for distinguishing between normal and anomalous instances. For these reasons, the capture and processing of these data are dramatically increasing in terms of 'volume' , 'velocity' and 'variety' , which are referred to as the phenomena of 'Big Data' [53]. The Big Data paradigm poses a continuous challenge in the use of network or host data sources for the design of an effective and scalable ADS.

Monitoring and analysing network traffic have attained growing importance for several reasons. Firstly, they increase visibility to the user, system and application traffic by gathering and analysing network flow records which also helps to track the bandwidth consumption of users and systems to ensure robust service delivery. Secondly, they identify performance bottlenecks and minimising non-business bandwidth consumption. Thirdly, there are advantages related to IDS technology, which are the tracking of network traffic using protocol analysis for recognising potential attack profiles, such as UDP spikes. Finally, they monitor network traffic, peer-to-peer protocols and URLs for a specific device or network to determine suspicious activities and unauthorised access [5].

In the literature, if data does not fit a normal distribution, it will be better to fit and detect outliers/anomalies using mixture models, especially Gaussian Mixture Model (GMM), Beta Mixture Model (BMM) or Dirichlet Mixture Model (DMM) [15, 17, 34, 50]. According to [17, 18, 21], the DMM can fit and define the boundaries of data better than other mixture models because it consists of a set of probability distributions. Moreover, the DMM is more suitable for modelling streaming data, for example, data originating from videos, images, or network traffic. The mathematical characteristics of the DMM also permit the representation of samples in a transformed space in which features are independent and identically distributed (i.i.d.). In the case of high dimensionality, the DMM for clustering data provides higher accuracy than other mixture models [9]. Therefore, we use this model to properly fit network data using the lower-upper Interquartile Range (IQR) [40] as a threshold to detect any observation outside them as an anomaly.

In this chapter, we propose a scalable framework for building an effective and lightweight ADS that can efficiently identify suspicious patterns over network systems. The framework consists of a capturing and logging module to sniff and record data, a pre-processing module to analyse and filter these data and the proposed ADS statistical decision engine, based on the DMM, for recognising abnormal behaviours in network systems. The DMM model is a statistical technique developed based on the method of anomaly detection which computes the density of Dirichlet distributions for the normal profile (i.e., the training phase) and testing phase (using the parameters estimated from the training phase). The decision-making method for identifying known and new anomalies is designed by specifying a threshold of the lower-upper IQR for the normal profile and considering any deviation from it as an attack.

The performance of this framework is evaluated on two well-known datasets, the NSL-KDD[1], which is an improved version of the KDD99 and the most popular dataset used for evaluating IDSs [48], and our UNSW-NB15[2] which involves a wide variety of contemporary normal, security and malware events [34]. The Dirichlet mixture model based anomaly detection technique is compared with three recent techniques, namely the Triangle Area Nearest Neighbours (TANN) [49], Euclidean Distance Map (EDM) [46] and Multivariate Correlation Analysis (MCA) [47]. These techniques were developed based on computing distances and correlations between legitimate and malicious vectors, which cannot often find a clear difference between these vectors, especially with modern attack styles that mimic normal ones.

---

[1]The NSLKDD dataset, https://web.archive.org/web/20150205070216/http://nsl.cs.unb.ca/NSL-KDD/, November 2016.

[2]The UNSW-NB15 dataset, https://www.unsw.adfa.edu.au/australian-centre-for-cyber-security/cybersecurity/ADFA-NB15-Datasets/, November 2016.

However, our technique was established using the methods of the Dirichlet mixture model and the accurate boundaries of interquartile range that properly find the small differences between these vectors, considerably improving the detection accuracy.

The key contributions of this chapter are as follows.

1. We propose a new scalable framework for anomaly detection system on large-scale networks. In this framework, we also develop a novel decision engine based on the Dirichlet Mixture Model and lower-upper Interquartile Range to efficiently detect malicious events.
2. We describe how statistical analysis can define the normality of network data to choose the proper model that correctly fits the data and make an intelligent decision-making method which discriminates between normal and abnormal observations.
3. A performance evaluation of this framework is conducted on two benchmark datasets: the NSL-KDD which is the most common dataset and UNSW-NB15 which is the latest dataset used for assessing IDSs, as well as comparing this technique with three existing techniques to assess its reliability for detecting intrusions.

The rest of this chapter is organised as follows. The background on Intrusion and Anomaly Detection Systems are presented in Sect. 2. Section 3 describes related work on decision engine approaches. The DMM-based technique is explained in Sect. 4 and the proposed scalable framework discussed in Sect. 5. The experimental results and discussions are provided in Sect. 6. Finally, concluding remarks are presented in Sect. 7.

2   Background on Intrusion and Anomaly Detection Systems

An Intrusion Detection System (IDS) is defined as a technique for monitoring host or network activities to detect possible threats by measuring their violations of Confidentiality, Integrity and Availability (CIA) principles [6, 42, 43]. There are two types of IDSs depending on the data source: a host-based IDS monitors the events of a host by collecting information about activities which occur in a computer system [51] while a network-based IDS monitors network traffic to identify remote attacks that take place across that network [11, 38].

IDS methodologies are classified into three major categories: Misuse-based (MDS); Stateful Protocol Analysis (SPA); and ADS [29, 38]. A MDS monitors host or network data audits to compare observed behaviours with those on an identified blacklist. Although it offers relatively high Detection Rates (DRs) and low FPRs, it cannot identify any zero-day attack. Also, it requires a huge effort to regularly update the blacklist which is a set of rules for each malicious category generated by security expertise [47]. A SPA inspects protocol states, especially a pair of request-response protocols, for example, HTTP ones. Although it is quite similar to an ADS, it depends on vendor-developed profiles for each protocol. Finally, an ADS establishes a normal profile from host or network data and discovers any variation from it as an attack. It can detect existing and new attacks, as it does not require any effort to generate rules, an ADS has become a better solution than MDS and SPA [13, 34, 38, 43]. However, it still has some challenges, as explained in Sect. 2.2, that we will try to mitigate.

An IDS's deployment architecture is classified as either distributed or centralised. A distributed IDS is a compound system involving several intrusion detection subsystems installed at different locations and connected in order to transfer relevant information. In contrast, a centralised IDS is a non-compound system deployed at only one location, with its architecture dependent on an organisation's size and sensitivity of its data which should be considered in terms of its deployment [28].

2.1 ADS Components

A typical ADS contains four components, a data source, a data pre-processing module, a decision engine technique and a defense response module [13, 34], as depicted in Fig. 1 and described in detail below.

- **Data source module**—This component is an essential part of any ADS that provides the potential host or network audit data to enable the DE to classify observations as either normal or attack [33]. Several data sources have been collected in offline datasets, such as the KDD CUP 99, NSL-KDD and UNSW-NB15, which consist of a wide variety of normal and malicious records for evaluating the performance of DE approaches. With the high speeds and large sizes of current communication systems, each data source has big data terms, i.e., a large volume, velocity and variety. Therefore, it is vital to design an effective and scalable ADS which can handle such data and make the correct decision upon the detection of malicious observations.

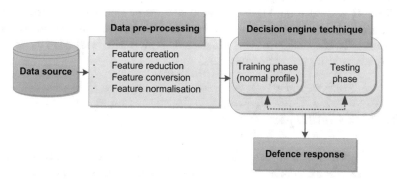

Fig. 1 Components of ADS

- **Data pre-processing module**—Data pre-processing is an important stage in any framework involving learning from data, as it handles and filters the input audit data by removing duplicated, irrelevant and/or noisy features to create a set of patterns that are then passed to the DE with the aim of improving the performance of that DE for detecting anomalous activity. It includes the following set of functions.

  - **Feature creation**—Feature creation constructs a set of features/attributes from host or network data using different tools, for instance, BRO-IDS, Netflow and Netmate. It is impossible to operate an IDS on raw data without mining a subset of features, such as the attributes in the NSL-KDD and UNSW-NB15 datasets.
  - **Feature reduction**—The function of feature reduction is to exclude unnecessary and duplicated attributes, and can be divided into feature selection and feature extraction methods. The first finds a subset of the original features and the second transforms the data from a high- to lower-dimensional space, such as using Principal Component Analysis (PCA) [12].
  - **Feature conversion**—The function of feature conversion is to convert feature types, which can be numeric or symbolic, into numeric values for ease of use in decision engine approaches as data analytics and statistical decision engine cannot use symbolic features to define data patterns.
  - **Feature normalisation**—Feature normalisation is a measure for scaling data features into a particular range, for example, [0,1], and is important for eliminating bias from raw data without modifying the statistical properties of the attributes.

- **DE module**—Intuitively, the decision engine is responsible for a critical stage, which is the design of an effective and efficient system for discovering intrusive activities in large-scale data in real time. Selecting the appropriate functions for a DE approach, and its training and testing phases, contributes to measuring the effectiveness of an IDS as, if it is not performed properly, the overall protection level will be easily compromised.
- **Security response module**—This module aims at indicating and explaining a decision taken by the system or administrators to prevent attack activities. More specifically, if a malicious event is detected, an alert will be raised to the security administrator for preventing this event.

2.2 ADS Challenges

Although a MDS cannot recognise zero-day attacks or even variants of existing attacks, it is still a common defence solution used in commercial products. On the contrary, an ADS can detect serious threats but has often been faced with potential challenges for its effective design. These challenges could be explored using an anomaly-based method, which is the construction of a purely normal profile with any variation from it declared an anomaly [24, 30, 37, 53], as follows.

- Establishing a profile which includes all possible normal patterns is very complex to achieve, as the boundary between normal and suspicious activities is always inaccurate. There are False Positive Rate (FPR) and False Negative Rate (FNR) errors which occur when a normal behaviour falls in an attack region (i.e. it is classified as an attack) and a malicious behaviour in a normal region, respectively.
- When designing the architecture of an adaptive and scalable ADS, it requires a careful analysis to discriminate attacks from the normal profile as sophisticated malicious activities, such as stealth and spy attacks [20], can adapt to be almost the same as normal patterns. Consequently, methods for detecting them have to analyse and inspect the potential characteristics of the network traffic.
- Real-time detection is also very challenging for reasons which increase its processing time and false alarm rate if not properly addressed. Firstly, the features created for network traffic may contain a set of noisy or irrelevant features. Secondly, the lightweight detection methods need to be carefully adopted, with respect to the above problems.
- Obtaining a decent-quality dataset is usually a major concern for evaluating, learning and validating ADS models. It should have a wide range of modern normal and malicious observations as well as being correctly labelled. This requires a huge effort of analysing the data in order to ensure establishing an authentic truth table, which has the security events and malware for the correct labelling process.
- The deployment of an ADS architecture is often difficult in large-scale environments, in particular, cloud computing and SCADA systems, have multiple nodes which could be either centralised or distributed. Also, the high speeds and a large amount of data transferring between these nodes often affect the performance of an ADS.

2.3   ADS Deployment in Enterprise Systems

With the new era of the Internet of Things (IoT), which is the networked interconnections of everyday objects often associated with their ubiquitous use, many applications and systems need to be protected against intrusive activities. As cloud computing environments and Supervisory Control and Data Acquisition (SCADA) systems are currently fully dependent on the Internet, they require an adaptable and scalable ADS for identifying the malicious events they frequently face. Cloud computing is a "network of networks" based on Internet services in which virtual shared servers provide the software, platform, infrastructure and other resources [3]. It consists of the three service models Software as a Service (SaaS), Platform as a Service (PaaS) and Infrastructure as a Service (IaaS) [30]. To detect attacks, an

IDS can be installed on a virtual server as a host-based IDS or deployed across the network as a network-based IDS or, by configuring both, provide a better defence.

SCADA systems monitor and control industrial and critical infrastructure functions, for example, water, electricity, railway, gas and traffic [16]. Like the cloud computing environment, with the rapid increase in the Internet and interconnected networks, these systems face complicated attacks, such as DoS and DDoS, which highlight the need for stronger SCADA security. However, designing the architecture and deployment of an adaptive and scalable ADS for these environments has become a big challenge because the high speeds and large sizes of existing networks generate a massive number of packets each time, and those packets should be inspected simultaneously in order to identify malicious activities.

3 Related Work on Decision Engine Approaches

Many researchers have investigated decision engine approaches, which can be categorised into five types: classification-based approaches [7, 10, 13]; clustering-based approaches [2, 7, 13, 17]; knowledge-based approaches [7, 10, 13]; combination-based approaches [7, 11, 13, 52]; and statistical-based approaches [7, 13, 34, 47], as illustrated in Table 1. Firstly, classification is a way of categorising data observations in particular classes in a training set with a testing set containing other instances for validating these classes; for instance, Horng et al. [23] proposed a Network-based ADS which included a hierarchical clustering and support vector machine to reduce the training time and improve detection accuracy. Ambusaidi et al. [4] developed a least-square support vector machine for the design of a lightweight Network-based ADS by selecting the significant features of network data and detecting anomalies. Recently, Dubey et al. [14] developed a Network-based ADS based on the collection techniques of an artificial neural network, k-means and Naïve-Bayes to improve the detection of malicious activities. However, overall, classification-based IDSs rely heavily on the assumption that each classifier has to be adjusted separately and always consume more resources than statistical techniques.

Secondly, clustering involves unsupervised machine-learning algorithms which allocate a set of data points to groups based on the similar characteristics of the points, for example, distance or probability measures. Nadiammai et al. [35] analysed and evaluated k-means, hierarchical and fuzzy c-means clustering techniques for building a Network-based ADS and reported that the complexity and detection accuracy of the fuzzy c-means algorithm were better than those of the others. Jadhav et al. [25] proposed a Network-based ADS based on clustering network packets and developed a new data pre-processing function using the fuzzy logic technique for classifying the severity of attacks in network traffic data. Zainaddin et al. [52] proposed a hybrid of fuzzy clustering and an artificial neural network to construct a Network-based ADS which efficiently detected malicious events. Clustering-based ADS techniques have several advantages. Firstly, they group data points in an unsupervised manner which means that they do not need to provide

Table 1 Comparison of decision engine approaches

Decision engine approaches	Related works	Advantages	Disadvantages
Classification	Horng et al. [23], Ambusaidi et al. [4], Dubey et al. [14]	• Provide higher detection rate and lower false positive rate if the network data is correctly labelled	• Depend on the assumption that each classifier has to be built separately • Consume more computational resources
Clustering	Nadiammai et al. [35], Jadhav et al. [25], Zainaddin et al. [52]	• Group data with no need to the class label • Decrease processing times	• Rely on the efficiency of building a normal profile • Require a higher time while updating this profile
Knowledge	Naldurg et al. [36], Hung et al. [24]	• Discriminate existing attacks • Provide higher detection rate	• Take too much time during the processing • Use static rules for defining malicious patterns
Combination	Perdisci et al. [37], Aburomman et al. [1], Shifflet [45]	• Achieve higher accuracy and detection • Demand only a set of controlling parameters to be adjusted	• Need a huge effort to integrate some techniques • Take a long processing time than other techniques
Statistics	Fan et al. [17], Greggio [18], Zhiyuan et al. [47]	• Accomplish higher accuracy and detection if a baseline of identifying attacks correctly adapted • Do not consume resources like other techniques	• Require accurate analysis to select the correct baseline • Need new functions to define attack types

class labels for observations. Secondly, they are effective for grouping large datasets into similar groups to detect network anomalies. However, in contrast, clustering is highly dependent on the efficacy of constructing a normal profile and the difficulty of automatically updating it.

Thirdly, knowledge-based methods establish a set of patterns from input data to classify data points with respect to class labels, with common knowledge-based ADSs rule-based, expert systems and ontologies. Naldurg et al. [36] suggested a framework for intrusion detection using temporal logic specifications with intrusion patterns formulated in a logic structure called EAGLE. It supported data values and parameters in recursive equations and enabled the identification of intrusions with temporal patterns. Hung et al. [24] presented an ontology-based approach for establishing a Network-based ADS according to the end-users' domain in which, as ontologies were applied as a conceptual modelling technique, a Network-based ADS could be simply built. Knowledge-based algorithms have some advantages. They are sufficiently robust and flexible to detect existing attacks in a small-scale system and achieve a high DR if a significant knowledge base about normal and abnormal instances can be correctly extracted. Conversely, they have FPRs due to the unavailability of biased normal and intrusion audit data and cannot identify rare or zero-day anomalies, and dynamically updating their rules is difficult.

Fourthly, combination-based techniques use many methods to effectively classify data instances, with most used for ADSs ensemble- and fusion-based techniques; for instance, Perdisci et al. [37] established a high-speed payload Network-based ADS based on an ensemble of one-class support vector machine for improving detection accuracy. Aburomman et al. [1] suggested an ensemble method which used PSO-generated weights to build a hybrid of more accurate classifiers for a Network-based ADS created based on local unimodal sampling and weighted majority algorithm approaches to improve the accuracy of detecting attacks. Shifflet [45] discussed a platform which enabled a hybrid of classification techniques to be executed together to build a fusion mechanism for the state of a network that was capable of efficiently detecting anomalous activities. Combination-based methods are advantageous as they achieve higher accuracy and detection rate than single ones while requiring a set of controlling parameters that can be easily adjusted. However, adopting a sub-set of consistent and unbiased classification techniques is difficult because it depends on using a hybridisation measure to combine them. Also, it is evident that their computational costs for large amounts of network traffic data are high due to the number of classifiers used.

Finally, in statistical-based approaches, an anomaly is a rare event which occurs among natural data ones and is measured by statistical methods which could be of the first order, such as means and standard deviations, second order, such as correlation measures, or third order, such as hypothesis testing and mixture models; for example, Fan et al. [17] developed an unsupervised statistical technique for identifying network intrusions in which legitimate and anomalous patterns were learned through finite generalised Dirichlet mixture models based on a Bayesian inference, with the parameters of the models and saliency of features simultaneously estimated. Greggio [18] designed a Network-based ADS based on the unsupervised fitting of network data using a Gaussian Mixture Model which selected the number of mixture components and fit the parameter for each component in a real environment. They extended their study to provide an efficient method for the varied learning of finite Dirichlet mixture models to design a Network-based ADS. This approach was based on the establishment and optimisation of a lower boundary for the likelihood of the model by adopting factored conditional distributions through its variables.

Overall, although ADS statistical-based approaches can analyse and determine the potential characteristics of normal and abnormal observations, identifying them and defining a certain baseline which distinguishes between normal and abnormal instances need accurate analysis. Therefore, we propose the methodology of the Dirichlet mixture model with the precise boundaries of interquartile range function as a decision engine. This is one of the statistical approaches that can define the inherent patterns of both legitimate and malicious features and observations, finding a clear variation between these observations. However, the other approaches often depend on many internal processes with a kernel function(s) that have to be adjusted for each problem. The main motivation for selecting this methodology is that statistical analytics of network data have shown that these data do not belong to a Gaussian distribution [17, 34]. Therefore, it is better to apply non-Gaussian distributions, such as Dirichlet mixture model to correctly fit network data using the lower-upper interquartile range function to detect any observation outside this range as an outlier/anomaly.

4 DMM-Based ADS Technique

This section describes the mathematical aspects of estimating and modelling data using the DMM, and discusses the proposed methodology for using this model to build an effective ADS.

4.1 Finite Dirichlet Mixture Model

Because a finite mixture model can be considered a convex combination of two or more Probability Density Functions (PDFs), the joint properties of which can approximate any arbitrary distribution, it is a powerful and flexible probabilistic modelling tool for handling multivariate data, such as network data [54]. A finite mixture of Dirichlet distributions with K components is shown in Fig. 2 and is given by [8, 18]

$$p(X|\pi, \alpha) = \sum_{i=1}^{K} \pi_i Dir(X|\alpha_i) \tag{1}$$

where $\pi = (\pi_1, \ldots, \pi_K)$ refers to the mixing coefficients, which are positive, with their summation 1, $\sum_{i=1}^{K} \pi_i, \alpha = (\alpha_1, \ldots, \alpha_K)$, and $Dir(X|\alpha_i)$ indicates the Dirichlet distribution of component i with its own positive parameters $(\alpha = (\alpha_{i1}, \ldots, \alpha_{iS}))$ as

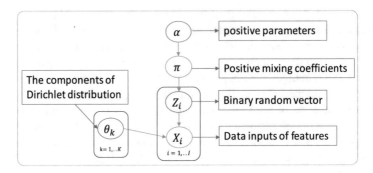

Fig. 2 Finite mixture model

$$Dir(X|\alpha_i) = \frac{\Gamma(\sum_{s=1}^{S} \alpha_{is})}{\prod_{s=1}^{S} \Gamma(\alpha_{is})} \prod_{s=1}^{S} X_s^{\alpha_{is}-1} \tag{2}$$

where $X = (X_1, \ldots, X_S)$, S is the dimensionality of X and $\sum_{s=1}^{S} x_s = 1, 0 \leq X_s \geq 1$ for $s = 1, \ldots, S$. It is worth noting that a Dirichlet distribution is used as a parent distribution to directly model the data rather than as a prior to the multinomial.

Considering a set of N independent identically distributed (*i.i.d*) vectors ($X = \{X_1, \ldots, X_N\}$) assumed to be produced from the mixture distribution in Eq. (1), the likelihood function of the DMM is

$$p(X|\pi, \alpha) = \prod_{l=1}^{N} \{\sum_{i=1}^{K} \Pi_i Dir(X_l|\alpha_i)\} \tag{3}$$

The finite mixture model in Eq. (1) is considered as a latent variable model. Therefore, for each vector (X_i), we introduce a K-dimensional binary random vector ($Zi = \{Z, \ldots, Z_{iK}\}$), where $Z_{is} \in \{0, 1\}$, $\sum_{i=1}^{K}$ and $Z_{is} = 1$ if X_i belongs to component i, otherwise 0. For the latent variables ($Z = \{Z_1, \ldots, Z_N\}$), which are actually hidden ones that do not appear explicitly in the model, the conditional distribution of Z given the mixing coefficients (π) is defined as

$$p(Z|\pi) = \prod_{l=1}^{N} \prod_{i=1}^{K} \pi_i^{Z_{li}} \tag{4}$$

Then, the likelihood function with the latent variables, which is actually the conditional distribution of a dataset L given the class labels, Z can be written as

$$p(X|\pi, \alpha) = \prod_{l=1}^{N} \prod_{i=1}^{K} Dir(X_l|\alpha_i) \tag{5}$$

Given a dataset L has a set of features D, an important problem is the learning process of the mixture parameters, that is, both estimating the parameters and selecting the number of components (K). In order to estimate these parameters and select the number of components correctly, we apply the Maximum Likelihood (ML) proposed in [30]. We suggest a new DMM for designing an ADS, namely DMM-ADS, in which includes training and testing phases for learning and validating network data. In the training phase, the DMM parameters and IQR are estimated to construct a normal profile, with abnormal instances identified in the testing phase, as detailed in the following two sections.

4.2   Training Phase of Normal Instances

The construction of a purely normal training set is extremely vital to ensure correct detection. Given a set of normal instances $(r_{1:n}^{normal})$ in which each record comprises a set of features D, where $r_{1:n}^{normal} = \{x_1, x_2, \ldots, x_D\}^{normal}$, the normal profile includes only statistical measures from $r_{1:n}^{normal}$. They include the estimated parameters (π, α, Z) of the DMM to compute the PDF of the Dirichlet distribution $(Dir(X|\pi, \alpha, Z))$ for each observation in the training set.

Algorithm 1 involves the proposed steps for establishing a normal profile (pro), with the parameters (π, α, Z) of the DMM computed for all the normal observations $(r_{1:n}^{normal})$ using the equations published in [30], and then the PDFs of the attributes $(X_{1:D})$ calculated using Eqs. (1)–(5). Next, the IQR is computed by subtracting the first quartile from the third quartile of the PDFs [40] to establish a threshold for recognising abnormal instances in the testing phase. It is acknowledged that quantiles are dividing a range of data into contiguous intervals with equal probabilities [40].

Algorithm 1: Normal profile construction of normal instances

    Input: normal instances $(r_{1:n}^{normal})$
    Output: normal profile (pro)
1: **for** each record i in $(r_{1:n}^{normal})$ **do**
2:     estimate the parameters (π_i, α_i, Z_i) of the DMM as in [29]
3:     calculate the PDFs using equations (1) to (5) based on the estimated parameters of Step 2
4: **end for**
5: compute $lower = quantile(PDFs, 1)$
6: compute $upper = quantile(PDFs, 3)$
7: compute $IQR = upper - lower$
8: pro $\leftarrow ((\pi, \alpha_i, Z_i), (lower, upper, IQR))$
9: **return** pro

Algorithm 2: Testing phase and decision-making method

    Input: observed instance ($r^{testing}$), pro
    Output: normal or attack
  1: calculate the PDFtesting using equations using the parameters (π_i, α_i, Z_i)
  2: **if** ($PDF^{testing} < (lower - w * (IQR)) \parallel (PDF^{testing} > (upper + w * (IQR))$ **then**
  3:    **return** attack
  4: **else**
  5:    **return** normal
  6: **end if**

4.3 Testing Phase and Decision-Making Method

In the testing phase, the Dirichlet PDF ($PDF^{testing}$) of each observed record ($r^{testing}$) is computed using the same parameters estimated for the normal profile ((π, α_i, Z_i),($lower, upper, IQR$)). Algorithm 2 includes the steps in the testing phase and decision-making method for identifying the Dirichlet PDFs of the attack records, with step 1 constructing the PDF of each observed record using the stored normal parameters (π_i, α_i, Z_i).

Steps 2 to 6 define the decision-making process. The IQR of the normal instances is computed to find the outliers/anomalies of any observed instance ($r^{testing}$) in the testing phase which are considered to be any observations falling below ($lower - w * (IQR)$) or above ($upper + w * (IQR)$), where w indicates the interval values between 1.5 and 3 [40]. The detection decision is based on considering any $PDF^{testing}$ falling out of this range as an attack record, otherwise normal.

5 Scalable ADS Framework

This section discusses a proposed scalable framework for developing an effective ADS which identifies malicious activities in large-scale environments. It consists of three modules, capturing and logging, data pre-processing and a DMM-based ADS technique, as shown in Fig. 3.

In the first phase, a set of attributes is created from network traffic to capture network connections for a particular time window. It is observed that the best way of analysing network data is to sniff the traffic from the router located at each network node and aggregate only relevant network packets [46, 47]. Secondly, the pre-processing step filters network data by converting symbolic features into numeric ones, as shown in Fig. 4.

The reason for this conversion process is because statistical methods, such as the DMM-based ADS technique, can handle only numeric data. Furthermore, selecting the most significant features to improve the performance and reduce the processing time of the decision engine in order to deploy it in real time. Finally, we propose the

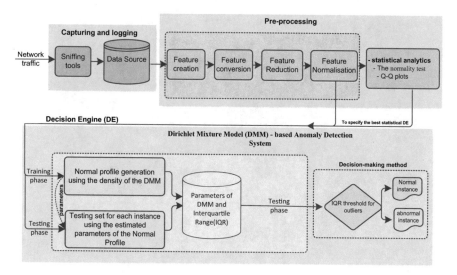

Fig. 3 Proposed scalable framework for designing effective ADS

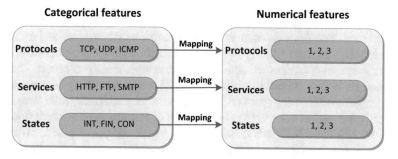

Fig. 4 Example of converting categorical features into numerical features using UNSW-NB15 dataset

DMM-based ADS technique as the decision engine, which efficiently discriminates between legitimate and suspicious instances, as discussed in the previous section.

5.1 Capturing and Logging Module

This module sniffs network data and stores them to be processed for the decision engine, like the steps for creating our UNSW-NB15 dataset [33, 34]. An IXIA PerfectStorm tool,[3] which has the capability to determine a wide range of network

---

[3]The IXIA Tools, https://www.ixiacom.com/products/perfectstorm, November 2016.

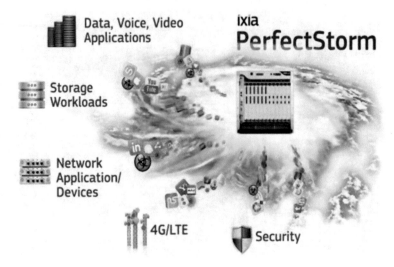

Fig. 5 Functions of IXIA PerfectStorm tool

segments and elicits traffic for several web applications, such as Facebook, Skype, YouTube and Google, was used to mimic recent realistic normal and abnormal network traffic, as shown in Fig. 5. It could also simulate the majority of security events and malicious scripts which is difficult to achieve using other tools. The configuration of the UNSW-NB15 testbed was used to simulate a large-scale network and a Tcpdump tool to sniff packets from the network's interface while Bro, Argus tools and other scripts were used to extract a set of features from network flows.

In [33, 34], the UNSW-NB15 was created, which comprises a wide variety of features. These features can be classified into packet-based and flow-based. The packet based features help in examining the packet payload and headers while the flow based features mine information from the packet headers, such as a packet direction, an inter-arrival time of packets, the number of source/destination IPs for a particular time window, and an inter-packet length. AS depicted in Fig. 6, the pcap files[4] of this dataset were processed by the BRO-IDS and Argus tools to mine the basic features. Then, we developed a new aggregator module to correlate its flows. These flows were aggregated for each 100 connections, where packets with the same source/destination IP addresses and ports, timestamp, and protocol were collected in a flow record [31, 32]. This module enables to establish monitoring applications for analysing network characteristics such as capacity, bandwidth, rare and normal events.

---

[4]**Pcap** refers to **p**acket **cap**ture, which contains an Application Programming Interface (API) for saving network data. The UNIX operating systems execute the pcap format using the libpcap library while the Windows operating systems utilise a port of libpcap, called WinPcap.

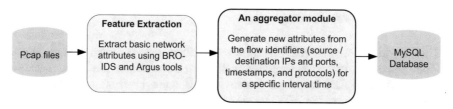

Fig. 6 Extracting and aggregating features of UNSW-NB15 dataset

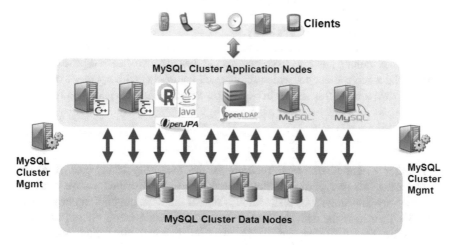

Fig. 7 Architecture of MySQL cluster CGE

These features were recorded using the MySQL Cluster CGE technology[5] that has a highly scalable and real-time database, enables a distributed architecture to read and write intensive workloads and is accessed via SQL or NoSQL APIs, as depicted in Fig. 7. It can also support memory-optimised and disk-based tables, automatic data partitioning with load balancing, and can add nodes to a running cluster for handling online big data. Although this technology has a similar architecture to Hadoop tools[6], which are the most popular for processing big offline data, an ADS has to detect malicious behaviours in real time. These features are then passed to the pre-processing module to be analysed and filtered.

---

[5]The MySQL Cluster CGE technology, https://www.mysql.com/products/cluster/, November 2016.

[6]The Hadoop technologies, http://hadoop.apache.org/, November 2016.

5.2 Pre-Processing Module

The pre-processing module determines and filters network data in four steps. Firstly, its feature conversion replaces symbolic features with numeric ones because our DMM-based ADS technique deals with only numeric attributes. Secondly, its feature reduction uses the PCA technique to adopt a small number of uncorrelated features. As the PCA technique is one of the best-known linear feature reduction techniques due to its the advantages. It requires less memory storage, having lower data transfer and processing times, as well as better detection accuracy than others [22, 26]. So, we chose it for this study.

Thirdly, feature normalisation arranges the value of each feature in a specific interval to eliminate any bias from raw data and easily visualise and process it. We applied the z-score function, which scales each feature (x) with a 0 mean (μ) and 1 standard deviation (δ), as shown in Fig. 8, to normalise the data using the formula

$$z = \frac{(x - \mu)}{\delta} \tag{6}$$

Another essential statistical measure is the normality test which is a way of assessing whether particular data follow a normal distribution. We used the Kolmogorov-Smirnov (K-S) test, which is one of the most popular, in our previous work [34]. In it, if the data do not follow a normal distribution, mixture models, such as the GMM, BMM and DMM, are used to efficiently define outliers. In this chapter, we use Q-Q plots to show that the network data do not follow a Gaussian distribution. A Q-Q plot is a graphical tool designed to draw two sets of

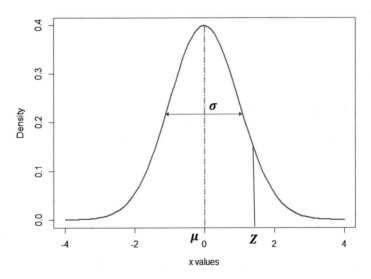

Fig. 8 Gaussian distribution with z-score parameters

quantiles against each other. If these sets are from the same distribution, the points form an almost straight line, with the others treated as outliers [19]. Therefore, it helps to track network flows and define which DE model is the best for identifying suspicious activities as outlier points, as shown in the results in Sect. 6.4. Overall, a statistical analysis is important for network data to make a decision regarding detecting and preventing malicious events.

6   Experimental Results and Discussions

This section discusses the datasets used for evaluating the proposed technique, and then the evaluation metrics applied for assessing the performance of the proposed technique compared with some peer techniques. Finally, the features selected from the NSL-KDD and UNSW-NB15 datasets, with the statistical results of these features are explained.

6.1   Datasets Used for Evaluation

Despite the NSL-KDD and KDD CUP 99 datasets being outdated and having several problems, in particular duplications of records and unbalancing of normal and attack records [33, 47, 48], they are widely used to evaluate NIDSs, due to a lack of accurate dataset availability. As most state-of-the-art detection techniques have been applied to these datasets, which are ultimately from the same network traffic, in order to provide a fair and reasonable evaluation of the performance of our proposed DMM-based ADS technique and comparison with those of related state-of-the-art detection approaches, we adopted the NSL-KDD dataset and contemporary UNSW-NB15 dataset which was recently released.

The NSL-KDD dataset is an improved version of the KDD CUP 99 dataset suggested by Tavallaee et al. in [48]. It addresses some of the problems in the KDD CUP 99 dataset, such as removing redundant records in the training and testing sets to eliminate any classifier being biased towards the most repeated records. Like in this dataset, in the NSL-KDD dataset, each record has 41 features and the class label. It consists of five different classes, one normal and four attack types (i.e., DoS, Probe, U2R and R2L), and includes two sets, training ('*KDDTrain$^+$ − FULL*' and '*KDDTrain$^+$ − 20%*') and testing ('*KDDTest$^+$ − 20%*' and '*KDDTest^{21} − newattacks*').

The UNSW-NB15 dataset has a hybrid of authentic contemporary normal and attack records. The volume of its network packets is approximately 100 GB with 2,540,044 observations logged in four CSV files. Each observation has 47 features and the class label which demonstrate its variety in terms of high dimensionality. Its velocity is, on average, 5–10 MB/s between sources and destinations which means higher data rate transmissions across the Ethernets which exactly mimic real network environments. The UNSW-NB15 dataset includes ten different classes, one normal and nine security and malware types (i.e., Analysis, Backdoors, DoS, Exploits, Generic, Fuzzers for anomalous behaviours, Reconnaissance, Shellcode and Worms) [33, 34].

6.2 Performance Evaluation

Several experiments were conducted on the two datasets to measure the performance and effectiveness of the proposed DMM-based ADS technique using external evaluation metrics, including the accuracy, DR and FPR which depend on the four terms true positive (TP), true negative (TN), false negative (FN) and false positive (FP). TP is the number of actual attack records classified as attacks, TN is the number of actual normal records classified as normal, FN is the number of actual attack records classified as normal and FP is the number of actual normal records classified as attacks. These metrics are defined as follows.

- The **accuracy** is the percentage of all normal and attack records correctly classified, that is,

$$accuracy = \frac{(TP + TN)}{(TP + TN + FP + FN)} \tag{7}$$

- The **Detection Rate (DR)** is the percentage of correctly detected attack records, that is,

$$DR = \frac{TP}{(TP + FN)} \tag{8}$$

- The **False Positive Rate (FPR)** is the percentage of incorrectly detected attack records, that is,

$$FPR = \frac{FP}{(FP + TN)} \tag{9}$$

6.3 Pre-Processing Phase and Description

The DMM-based ADS technique was evaluated using the eight features from the NSL-KDD and UNSW-NB15 datasets adopted using the PCA listed in Table 2.

The proposed DMM-ADS technique was developed using the 'R language' on Linux Ubuntu 14.04 with 16 GB RAM and an i7 CPU processor. To conduct the experiments on each dataset, we selected random samples from the 'full' NSL-KDD dataset and the CSV files of the UNSW-NB15 dataset with various sample sizes

Table 2 Attributes selected from two datasets

Selected attributes	Description
NSL-KDD dataset	
srv_count	Number of connections to the same service as the current connection in the past 2 s
dst_host_srv_count	Number of connections to the same service in the past 100 connections
count	Number of connections to the same host as the current connection in the past 2 s
dst_host_same_srv rate	Number of connections to different service as the current connection in the past 2 s
dst_host_count	Number of connections to the same host in the past 100 connections
hot	Hot indicators, e.g., access to system directories, creation, and execution of programs
srv_diff_host_rate	Percentage of same service connections to different hosts
rerror_rate	Percentage of same host connections that have "REJ" errors
UNSW-NB15 dataset	
ct_dst_sport_ltm	Number of connections containing the same destination address and source port in 100 connections
tcprtt	Round-trip time of TCP connection setup computed by the sum of 'synack' and 'ackdat'
dwin	Value of destination TCP window advertisement
ct_src_dport_ltm	Number of connections containing the same source address and destination port in 100 connections
ct_dst_src_ltm	Number of connections containing the same source and destination address in 100 connections
ct_dst_ltm	Number of connections containing the same destination address in 100 connections
smean	Mean of flow packet sizes transmitted from source
service	Service types, e.g., HTTP, FTP, SMTP, SSH, DNS and IRC

between 80,000 and 200,000. In each sample, normal instances were approximately 60–70% of the total size, with some used to create the normal profile and the rest for the testing set.

6.4 Statistical Analysis and Decision Support

Statistical analysis supports the decisions of defining the type of modelling, which efficiently fits data to recognise outliers as attacks. As previously mentioned, the Q-Q plot is a graphical tool to check if a set of data come from a normal theoretical distribution. features are considered from a normal distribution if the values of those features fall on the same theoretical distribution line. Figure 9 represents that the

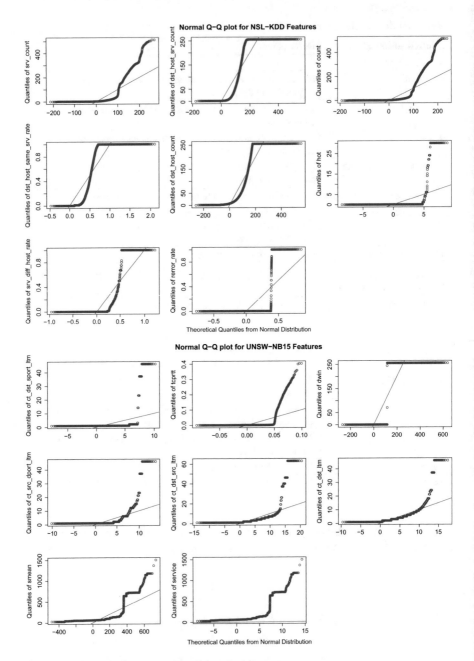

Fig. 9 Q-Q plot of the features selected from both datasets

selected features do not fall on the theoretical distribution lines (i.e., red ones), and there are much greater variations than the lines of feature values. We, therefore, decided to choose the DMM, as one of the best non-normal distribution, for fitting these features to build an ADS based on detecting the too far points of the feature values as anomalies.

The PDFs of the DMM are estimated for normal and abnormal instances using the NSL-KDD and UNSW-NB15 datasets to demonstrate to what extent these instances vary, as presented in Fig. 10. In the NSL-KDD dataset, the PDFs of the normal values range between 0 and 0.20, and their values are between -50 and 0. In contrast, the PDFs of the abnormal values falls between 0 and 0.5, and their values are between -30 and 0. As a result, it is noted that the PDFs of the normal instances are different from the attack ones. Likewise, in the UNSW-NB15 dataset, the PDFs of the normal instances are also dissimilar to the attack instances. These results assert that the proposed decision-making method in Algorithm 2 can effectively detect attacks due to the differences in their PDFs.

6.5   Performance of DMM-Based ADS Technique

The performance evaluation of the DMM-based ADS technique was conducted on the features selected from the two datasets, with the overall DR, accuracy and FPR values listed in Table 3. Figure 11 represents the Receiver Operating Characteristics (ROC) curves which display the relationship between the DRs and FPRs using the w values. It can be seen that the steady increase in the w value between 1.5 and 3 increased the overall DR and accuracy while decreasing the overall FPR.

In the NSL-KDD dataset, when the w value surged steadily from 1.5 to 3, the overall DR and accuracy increased from 93.1% to 97.8% and 93.2% to 97.8%, respectively, while the overall FPR reduced from 3.1% to 2.5%. Likewise, in the UNSW-NB15 dataset, the overall DR and accuracy increased from 84.1% to 93.9% and 89.1% and 94.3%, respectively, but the overall FPR reduced from 9.2% to 5.8% when the w value increased from 1.5 to 3.

Tables 4 and 5 show comparisons of the DRs of the record types for the w values on the NSL-KDD and UNSW-N15 datasets, respectively, which refers that, when the w value increased, the DR gradually improved. It is clear in Table 4 that the DMM-based ADS technique could detect the majority of record types of the NSL-KDD dataset with a normal DR varying between 96.7% and 99.8%, and the lowest FN rate when the w value changed from 1.5 to 3. Similarly, the DRs of the attack types increased gradually from an average of 93.2% to an average of 97.1%.

Table 5 indicates that the DMM-based ADS technique detected record types of the UNSW-NB15 dataset with normal DRs varying from 83.4% to 94.2% when the w value increased from 1.5 to 3. Similarly, the DRs of the attack types increased gradually from an average of 77.5% to an average of 93.2%.

The Shellcode, Fuzzers, Reconnaissance, and Backdoor attacks do not achieve the best DRs with the highest w, whereas the DRs of the other attacks, DoS, Generic,

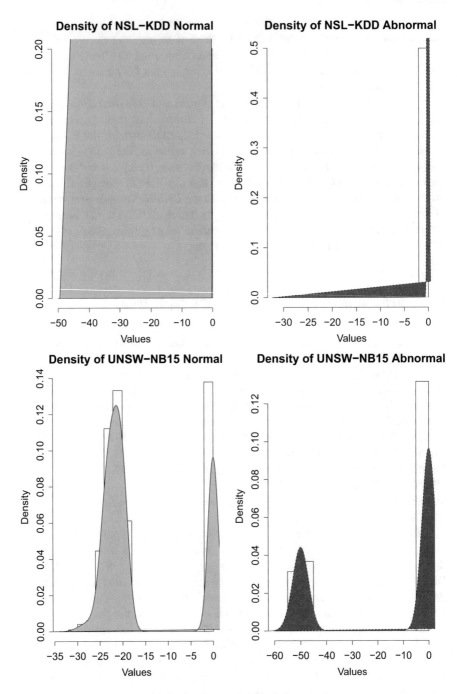

Fig. 10 Normal and abnormal PDFs from a sample of both datasets

Table 3 Performance evaluation of features selected from both datasets

| w value | Datasets | | | | | |
| | NSL-KDD | | | UNSW-NB15 | | |
	Detection rate	Accuracy	False positive rate	Detection rate	Accuracy	False positive rate
1.5	93.1	93.2	3.1	84.1	89.0	9.2
2	93.2	93.8	4.2	87.1	88.1	6.6
2.5	97.3	97.8	2.8	89.2	90.7	7.0
3	97.8	97. 8	2.5	93.9	94.3	5.8

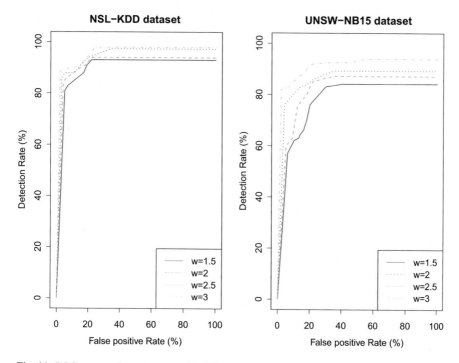

Fig. 11 ROC curves of two datasets with different w values

Exploits, and Worms, are lower, due to the slight similarities between these attack instances and normal ones. It can be noted that, as the variances of the selected features for these instances are close, the PDFs fell into each other in terms of decision-making.

Table 4 Comparison of detection rate (%) on NSL-KDD dataset

Instance type	w values			
	1.5	2	2.5	3
Normal	96.7	97.2	97.3	99.8
DoS	97.0	98.0	98.8	99.7
Probe	92.6	93.7	95.3	97.8
R2L	95.1	93.1	95.1	95.8
U2R	92.6	90.8	93.2	94.0

Table 5 Comparison of detection rate (%) on UNSW-NB15 dataset

Instance type	w values			
	1.5	2	2.5	3
Normal	83.4	83.0	89.7	94.2
DoS	89.1	89.1	88.2	88.1
Backdoor	63.5	72.2	74.2	71.3
Exploits	42.3	78.2	82.1	81.0
Analysis	73.8	76.3	78.0	81.1
Generic	78.1	89.4	88.1	87.7
Fuzzers	43.1	49.1	50.8	52.8
Shellcode	42.2	51.1	52.2	52.2
Reconnaissance	56.1	54.1	57.1	57.2
Worms	37.4	45.2	44.3	48.3

Table 6 Comparison of performances of four techniques

Technique	Detection rate (%)	False positive rate (%)
TANN [49]	91.1	9.4
EDM [46]	94.2	7.2
MCA [47]	96.2	4.9
DMM-based ADS	97.2	2.4

6.6   Comparative Study

The performance evaluation results for the DMM-based ADS technique based on the NSL-KDD dataset were compared with those from other three existing techniques, namely the Triangle Area Nearest Neighbours (TANN) [49], Euclidean Distance Map (EDM) [46] and Multivariate Correlation Analysis (MCA) [47], with their overall DRs and FPRs listed in Table 6. These techniques are used for comparing with our technique because they are the recent ones which have similar statistical measures to our DMM-based ADS. The DRs of The TANN, EDM and MCA were 91.1%, 94.2% and 96.2%, respectively, and their FARs 9.4%, 7.2% and 4.9%, respectively. In contrast, the DMM-based ADS achieved better results of a 97.2% DR and 2.4% FPR.

The key reason for the DMM-based ADS technique performing better than the other techniques was that the DMM fits the boundaries of each feature perfectly, because it has a set of distributions for computing the PDF of each instance. Moreover, the lower-upper IQR method could effectively specify the boundary between normal and outlier instances. However, despite the DMM-based ADS technique achieving the highest DR and lowest FPR on the NSL-KDD dataset, its performance on the UNSW-NB15 dataset was relatively lower due to slight variations between the normal and abnormal instances. This indicated the complicated patterns of contemporary attacks that almost mimic normal patterns.

6.7   Advantages and Disadvantages of DMM-Based ADS

The DMM-based ADS has several advantages. To start with, it is easily deployed on large-scale systems to detect malicious activity in real-time because its training and testing phases depend only on the DMM parameters of the normal profile. Since the decision-making method is used the lower-upper IQR rule as a threshold, it can identify the class label of each record with no dependency on other records. Moreover, the ease of updating the normal profile parameters, with respect to choose the best threshold. In contrast, if there are higher similarities between features, it will produce higher FPR, so we applied the PCA to reduce the number of features with selecting the highest variation of features for improving the performance of the proposed technique. Also, the DMM-based ADS cannot define attack types, such DoS and backdoors, as it was designed for handling binary classification (i.e., normal or attacks). For addressing this limitation, we will design a new statistical function to identify the PDF values of each attack type.

7   Conclusion

This chapter discussed a proposed scalable framework consisting of three main modules, namely, capturing and logging, pre-processing and a statistical decision engine. The purpose of the first module was to sniff and collect network data from a distributed database to easily handle large-scale environments while the second analysed and filtered network data to improve the performance of the decision engine. Finally, the third, the Dirichlet mixture model-based anomaly detection, was designed based on an anomaly detection methodology for recognising abnormal data using a lower-upper interquartile range as a decision-making method. The empirical results showed that a statistical analysis, such as Q-Q plots, helped to make a decision regarding choosing the best model for identifying attacks as outliers. The performance evaluation of the Dirichlet mixture model-based anomaly detection demonstrated that it was more accurate than some other significant methods. In future, we will investigate other statistical methods, such as a particle filter, with the aim of integrating them with the Q-Q plots to design a visual application for analysing and monitoring network data, and making decisions regarding specific intrusions. We will also extend this study to apply the architecture of the proposed framework in cloud computing and SCADA systems.

References

1. Aburomman, A.A., Reaz, M.B.I.: A novel svm-knn-pso ensemble method for intrusion detection system. Applied Soft Computing **38**, 360–372 (2016)
2. Ahmed, M., Mahmood, A.N., Hu, J.: A survey of network anomaly detection techniques. Journal of Network and Computer Applications **60**, 19–31 (2016)
3. Alqahtani, S.M., Al Balushi, M., John, R.: An intelligent intrusion detection system for cloud computing (sidscc). In: Computational Science and Computational Intelligence (CSCI), 2014 International Conference on, vol. 2, pp. 135–141. IEEE (2014)
4. Ambusaidi, M., He, X., Nanda, P., Tan, Z.: Building an intrusion detection system using a filter-based feature selection algorithm (2016)
5. traffic analysis, N.: Network traffic analysis (November 2016). URL https://www.ipswitch.com/solutions/network-traffic-analysis
6. Berthier, R., Sanders, W.H., Khurana, H.: Intrusion detection for advanced metering infrastructures: Requirements and architectural directions. In: Smart Grid Communications (SmartGridComm), 2010 First IEEE International Conference on, pp. 350–355. IEEE (2010)
7. Bhuyan, M.H., Bhattacharyya, D.K., Kalita, J.K.: Network anomaly detection: methods, systems and tools. IEEE Communications Surveys & Tutorials **16**(1), 303–336 (2014)
8. Bouguila, N., Ziou, D., Vaillancourt, J.: Unsupervised learning of a finite mixture model based on the dirichlet distribution and its application. IEEE Transactions on Image Processing **13**(11), 1533–1543 (2004)
9. Boutemedjet, S., Bouguila, N., Ziou, D.: A hybrid feature extraction selection approach for high-dimensional non-gaussian data clustering. IEEE Transactions on Pattern Analysis and Machine Intelligence **31**(8), 1429–1443 (2009)
10. Chandola, V., Banerjee, A., Kumar, V.: Anomaly detection: A survey. ACM computing surveys (CSUR) **41**(3), 15 (2009)
11. Corona, I., Giacinto, G., Roli, F.: Adversarial attacks against intrusion detection systems: Taxonomy, solutions and open issues. Information Sciences **239**, 201–225 (2013)
12. Ding, Q., Kolaczyk, E.D.: A compressed pca subspace method for anomaly detection in high-dimensional data. IEEE Transactions on Information Theory **59**(11), 7419–7433 (2013)
13. Dua, S., Du, X.: Data mining and machine learning in cybersecurity. CRC press (2016)
14. Dubey, S., Dubey, J.: Kbb: A hybrid method for intrusion detection. In: Computer, Communication and Control (IC4), 2015 International Conference on, pp. 1–6. IEEE (2015)
15. Escobar, M.D., West, M.: Bayesian density estimation and inference using mixtures. Journal of the american statistical association **90**(430), 577–588 (1995)
16. Fahad, A., Tari, Z., Almalawi, A., Goscinski, A., Khalil, I., Mahmood, A.: Ppfscada: Privacy preserving framework for scada data publishing. Future Generation Computer Systems **37**, 496–511 (2014)
17. Fan, W., Bouguila, N., Ziou, D.: Unsupervised anomaly intrusion detection via localized bayesian feature selection. In: 2011 IEEE 11th International Conference on Data Mining, pp. 1032–1037. IEEE (2011)
18. Fan, W., Bouguila, N., Ziou, D.: Variational learning for finite dirichlet mixture models and applications. IEEE transactions on neural networks and learning systems **23**(5), 762–774 (2012)

19. Ghasemi, A., Zahediasl, S., et al.: Normality tests for statistical analysis: a guide for non-statisticians. International journal of endocrinology and metabolism **10**(2), 486–489 (2012)
20. Giannetsos, T., Dimitriou, T.: Spy-sense: spyware tool for executing stealthy exploits against sensor networks. In: Proceedings of the 2nd ACM workshop on Hot topics on wireless network security and privacy, pp. 7–12. ACM (2013)
21. Greggio, N.: Learning anomalies in idss by means of multivariate finite mixture models. In: Advanced Information Networking and Applications (AINA), 2013 IEEE 27th International Conference on, pp. 251–258. IEEE (2013)
22. Harrou, F., Kadri, F., Chaabane, S., Tahon, C., Sun, Y.: Improved principal component analysis for anomaly detection: Application to an emergency department. Computers & Industrial Engineering **88**, 63–77 (2015)
23. Horng, S.J., Su, M.Y., Chen, Y.H., Kao, T.W., Chen, R.J., Lai, J.L., Perkasa, C.D.: A novel intrusion detection system based on hierarchical clustering and support vector machines. Expert systems with Applications **38**(1), 306–313 (2011)
24. Hung, S.S., Liu, D.S.M.: A user-oriented ontology-based approach for network intrusion detection. Computer Standards & Interfaces **30**(1), 78–88 (2008)
25. Jadhav, A., Jadhav, A., Jadhav, P., Kulkarni, P.: A novel approach for the design of network intrusion detection system (nids). In: Sensor Network Security Technology and Privacy Communication System (SNS & PCS), 2013 International Conference on, pp. 22–27. IEEE (2013)
26. Lee, Y.J., Yeh, Y.R., Wang, Y.C.F.: Anomaly detection via online oversampling principal component analysis. IEEE Transactions on Knowledge and Data Engineering **25**(7), 1460–1470 (2013)
27. Li, W., Mahadevan, V., Vasconcelos, N.: Anomaly detection and localization in crowded scenes. IEEE transactions on pattern analysis and machine intelligence **36**(1), 18–32 (2014)
28. Milenkoski, A., Vieira, M., Kounev, S., Avritzer, A., Payne, B.D.: Evaluating computer intrusion detection systems: A survey of common practices. ACM Computing Surveys (CSUR) **48**(1), 12 (2015)
29. Minka, T.: Estimating a dirichlet distribution (2000)
30. Modi, C., Patel, D., Borisaniya, B., Patel, H., Patel, A., Rajarajan, M.: A survey of intrusion detection techniques in cloud. Journal of Network and Computer Applications **36**(1), 42–57 (2013)
31. Moustafa, N., Slay, J.: A hybrid feature selection for network intrusion detection systems: Central points. In: the Proceedings of the 16th Australian Information Warfare Conference, Edith Cowan University, Joondalup Campus, Perth, Western Australia, pp. 5–13. Security Research Institute, Edith Cowan University (2015)
32. Moustafa, N., Slay, J.: The significant features of the unsw-nb15 and the kdd99 data sets for network intrusion detection systems. In: Building Analysis Datasets and Gathering Experience Returns for Security (BADGERS), 2015 4th International Workshop on, pp. 25–31. IEEE (2015)
33. Moustafa, N., Slay, J.: Unsw-nb15: a comprehensive data set for network intrusion detection systems (unsw-nb15 network data set). In: Military Communications and Information Systems Conference (MilCIS), 2015, pp. 1–6. IEEE (2015)
34. Moustafa, N., Slay, J.: The evaluation of network anomaly detection systems: Statistical analysis of the unsw-nb15 data set and the comparison with the kdd99 data set. Information Security Journal: A Global Perspective (2016)
35. Nadiammai, G., Hemalatha, M.: An evaluation of clustering technique over intrusion detection system. In: Proceedings of the International Conference on Advances in Computing, Communications and Informatics, pp. 1054–1060. ACM (2012)
36. Naldurg, P., Sen, K., Thati, P.: A temporal logic based framework for intrusion detection. In: International Conference on Formal Techniques for Networked and Distributed Systems, pp. 359–376. Springer (2004)

37. Perdisci, R., Gu, G., Lee, W.: Using an ensemble of one-class svm classifiers to harden payload-based anomaly detection systems. In: Sixth International Conference on Data Mining (ICDM'06), pp. 488–498. IEEE (2006)
38. Pontarelli, S., Bianchi, G., Teofili, S.: Traffic-aware design of a high-speed fpga network intrusion detection system. IEEE Transactions on Computers 62(11), 2322–2334 (2013)
39. Ranshous, S., Shen, S., Koutra, D., Harenberg, S., Faloutsos, C., Samatova, N.F.: Anomaly detection in dynamic networks: a survey. Wiley Interdisciplinary Reviews: Computational Statistics 7(3), 223–247 (2015)
40. Rousseeuw, P.J., Hubert, M.: Robust statistics for outlier detection. Wiley Interdisciplinary Reviews: Data Mining and Knowledge Discovery 1(1), 73–79 (2011)
41. Saligrama, V., Chen, Z.: Video anomaly detection based on local statistical aggregates. In: Computer Vision and Pattern Recognition (CVPR), 2012 IEEE Conference on, pp. 2112–2119. IEEE (2012)
42. Seeberg, V.E., Petrovic, S.: A new classification scheme for anonymization of real data used in ids benchmarking. In: Availability, Reliability and Security, 2007. ARES 2007. The Second International Conference on, pp. 385–390. IEEE (2007)
43. Shameli-Sendi, A., Cheriet, M., Hamou-Lhadj, A.: Taxonomy of intrusion risk assessment and response system. Computers & Security 45, 1–16 (2014)
44. Sheikhan, M., Jadidi, Z.: Flow-based anomaly detection in high-speed links using modified gsa-optimized neural network. Neural Computing and Applications 24(3–4), 599–611 (2014)
45. Shifflet, J.: A technique independent fusion model for network intrusion detection. In: Proceedings of the Midstates Conference on Undergraduate Research in Computer Science and Mat hematics, vol. 3, pp. 1–3. Citeseer (2005)
46. Tan, Z., Jamdagni, A., He, X., Nanda, P., Liu, R.P.: Denial-of-service attack detection based on multivariate correlation analysis. In: International Conference on Neural Information Processing, pp. 756–765. Springer (2011)
47. Tan, Z., Jamdagni, A., He, X., Nanda, P., Liu, R.P.: A system for denial-of-service attack detection based on multivariate correlation analysis. IEEE transactions on parallel and distributed systems 25(2), 447–456 (2014)
48. Tavallaee, M., Bagheri, E., Lu, W., Ghorbani, A.A.: A detailed analysis of the kdd cup 99 data set. In: Proceedings of the Second IEEE Symposium on Computational Intelligence for Security and Defence Applications 2009 (2009)
49. Tsai, C.F., Lin, C.Y.: A triangle area based nearest neighbors approach to intrusion detection. Pattern recognition 43(1), 222–229 (2010)
50. Wagle, B.: Multivariate beta distribution and a test for multivariate normality. Journal of the Royal Statistical Society. Series B (Methodological) pp. 511–516 (1968)
51. Wu, S.X., Banzhaf, W.: The use of computational intelligence in intrusion detection systems: A review. Applied Soft Computing 10(1), 1–35 (2010)
52. Zainaddin, D.A.A., Hanapi, Z.M.: Hybrid of fuzzy clustering neural network over nsl dataset for intrusion detection system. Journal of Computer Science 9(3), 391 (2013)
53. Zuech, R., Khoshgoftaar, T.M., Wald, R.: Intrusion detection and big heterogeneous data: a survey. Journal of Big Data 2(1), 1 (2015)

Security of Online Examinations

Yousef W. Sabbah

Abstract Online-examination modeling has been advancing at a slow, thus steady pace. Such an endeavor is embedded in many of today's fast-paced educational institutions. So, the online examination (i.e. e-Examination) model demonstrated in this chapter proposes two major schemes that utilize the most up-to-date features of information and communication technology (ICT). We have integrated authentication methods into this model in the form of simulated and controlled, thus measurable enhancements. The new model complies with international examination standards and have been proved to be equally, if not more, immuned to its predecessor models, including classroom-based examination sessions. Therefore, it can be selected as a new model of examination to cut-down on the cost of exam administration and proctoring.

e-Examination systems are vulnerable to cyberattacks, leading to denial-of-service and/or unauthorized access to sensitive information. In order to prevent such attacks and impersonation threats, we have employed smart techniques of continuous authentication. Therefore, we propose two schemes; Interactive and Secure E-Examination Unit (ISEEU) which is based on video monitoring, and Smart Approach for Bimodal Biometrics Authentication in Home-exams (SABBAH) which implements bimodal biometrics and video-matching algorithms. Still, the model is scalable and upgradable to keep it open to smarter integration of state-of-the-art in the field of continuous authentication. For validation purposes, we have conducted a comprehensive risk analysis, and results show that our proposed model achieved higher scores than the previous ones.

1 Introduction

e-Learning utilizes Information and Communication Technology (ICT) to enhance the educational process. It is a modern model that provides an interactive-learning environment, which consists of tutors, students, contents, classrooms and the

Y.W. Sabbah (✉)
Faculty of Technology and Applied Sciences, Quality Assurance Department, Al-Quds Open University, Ramallah, Palestine
e-mail: ysabbah@qou.edu

© Springer International Publishing AG 2017
I. Palomares et al. (eds.), *Data Analytics and Decision Support for Cybersecurity*,
Data Analytics, DOI 10.1007/978-3-319-59439-2_6

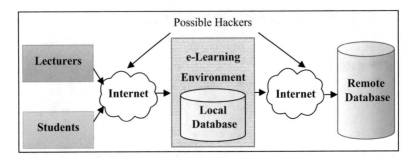

Fig. 1 Possible hackers in e-Learning environment [1] (Adapted)

educational process itself [1–8]. As depicted in Fig. 1, the Internet connects a learner with his lecturer and content regardless of place and time. Above and beyond, many academic institutions consider e-Learning a vital element of their information systems [1].

An e-Learning platform has different names, such as Learning and Course Management System (LCMS), Virtual Learning Environment (VLE), e-Learning Portal, etc. For instance, Moodle is a LCMS that supports social constructive pedagogy with interactive style. Interactive material includes Assignments, Choices, Journals, Lessons, Quizzes and Surveys [9]. Accordingly, we have implemented our proposed model based on Quiz module in Moodle.

For better performance, an e-Learning platform should be integrated withuniversity management information system (UMIS). This integrated solution combines all relevant technologies in a single educational environment (i.e. an e-University) that provides students, instructors and faculties with all required services [10].

In this chapter, we propose an integrated e-learning solution that consists of main and complementary components, as shown in Fig. 2. The academic portal (AP), with single sign-on (SSO), represents the core of this solution, and the authoring tools are used to build its Content. The middle annulus represents the main components, which are classified, in the inner annulus, into three groups; e-Content, delivery and assessment. The outer annulus illustrates the complementary components that represent together a University Management Information System (UMIS). All components are interconnected and can exchange data through the AP.

At the beginning, e-Learning systems were treated as research projects that concentrate on functionalities and management tools rather than security [11]. Nowadays, these systems are operational and heavily used worldwide. In addition, possible hackers may be located in the connections between either users and the system or the system and remote database [1], as shown in Fig. 1. Therefore, e-Learning systems should be provided with sufficient security to ensure confidentiality, integrity, availability and high performance.

Moreover, e-Examination security occupies the highest priority in e-Learning solutions, since this module contains the most sensitive data. In addition, an

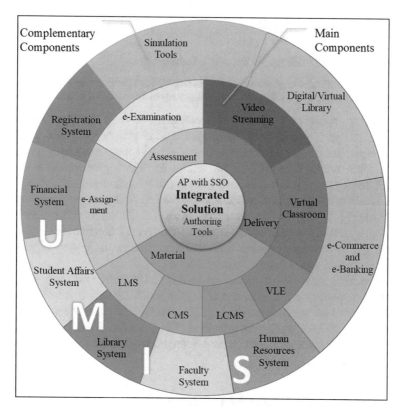

Fig. 2 A proposed integrated e-Learning solution; main and complementary components

efficient authentication method is required to make sure that the right student (e.g. examinee) is conducting an exam throughout the exam's period. The absence of trusted techniques for examinees' authentication is a vital obstacle facing e-Learning developers [12]. This is why opponents claim that e-Learning cannot provide a comprehensive learning environment, especially cheating-free online exams. Our contribution is to find a solution for this problem through a novel model for continuous authentication. This chapter introduces our proposed model in two schemes called ISEEU and SABBAH.

The chapter consists of four sections. The current section provides an overview of the main concepts of e-Learning and e-Examination systems and possible attacks that should be considered. The second section discusses the main security issues and authentication methods in e-Examination systems, as well as classification of the main existing authentication schemes. The third section describes our proposed continuous-authentication schemes. It provides a proposed implementation environment and settings and a comprehensive analysis, design and implementation of the schemes. Finally, the fourth section provides a comprehensive risk-analysis and evaluation to compare the suggested schemes with their predecessors, a full discussion of the results and conclusion, as well as challenges and future work.

2   e-Examination Security

Computer security is defined, in the NIST Handbook, as [13]:

> The protection afforded to an automated information system in order to attain the applicable
> objectives of preserving the integrity, availability, and confidentiality of its resources.

Computer and network security is a collection of tools and measures, which protect data that is stored in or exchanged among computers within a network [14]. Despite the major advancement introduced in system security in the last decade, many information systems fall victims to various cyberattacks worldwide [15, 16]. Information security is a serious issue and a fundamental requirement in e-Examination systems, since they add new threats and risks compared to other web-applications [7, 17–20]. In this context, security is discussed from two viewpoints; end-users (e.g. authors, teachers and students) and security functions (e.g. protecting content, personal security, access control, authentication and cryptography) [11]. Moreover, privacy issues should be investigated [19].

This section consists of three subsections. The first discusses the main e-Examination-security concepts that can assist in understanding the new features of our proposed authentication schemes. The second introduces the authentication methods that can be used in e-Examination. The last subsection describes the five existing authentication schemes that have been proposed in the previous studies.

2.1   e-Examination Security Concepts

This subsection aims at introducing the main e-Examination security concepts. We present security countermeasures and controls, C-I-A vs. P-I-A security goals, the need for continuous authentication and the impersonation threat types.

2.1.1   Security Countermeasures

Technical countermeasures and procedural requirements should be applied in e-Examination to ensure that lecturers, students and data are well-protected against possible risks [1, 21].

In general, *four technical security-countermeasures are used to measure security of e-Examination systems* or any computer system; confidentiality, integrity, availability and authenticity [1, 5, 14, 19, 22]. Regarding *procedural security-countermeasures, four essential requirements should be enforced*; security governance, security policy, security risk-management plan, and security-measures monitoring [1].

2.1.2 Security Controls

The following security controls are essential to protect e-Learning, thus e-Examination systems [1, 2, 14, 20, 23]:

- Access Control: Only the authorized entities can access a system.
- Encryption: Protection of private data against disclosure using cryptography.
- Firewalls: Filtering data exchanged between internal and external networks.
- Intrusion Detection: detection of attack attempts and alarm generation.
- Protection against Viruses and Spyware.
- Digital Signature: Ensuring that the received content is from a specific user.
- Digital Certificate: Verifying whether the transmitted digital content is genuine.
- Content Filter: Prevention of authorized entities from posting undesired content.

2.1.3 The Need for Continuous Authentication

More caution should be taken in online examination, where e-Examination systems should verify an examinee is the actual student [3, 21, 24–28]. Therefore, continuous and/or random authentication is required [21, 24, 25, 27, 28]. Confidentiality, integrity and availability (C-I-A) security goals can protect any system's hardware, software and data-assets against potential threats such as interception, modification, interruption and fabrication [25]. If the C-I-A goals are compromised, the critical assets will be compromised.

Apampa [25, 29] proposed that C-I-A security goals are unsuitable to protect all e-Examination assets, especially human assets (e.g. students, teachers and system administrators), since they have unpredictable attributes. For instance, the people who maintain the e-Examination system (i.e. system administrators) are valuable assets that do not depend on C-I-A goals; instead they should satisfy other goals. Students represent another example of such assets. To overcome this issue, Apampa [25, 29] proposed three new goals; Presence-Identity-Authentication (P-I-A), as indicated in Fig. 3 and defined below [25, 28]:

- *Presence* and continuously *authenticated presence*, which specifies a student's place.
- *Identity*, which differentiates a student from another.
- *Authentication*, which proves student's claimed identity.

2.1.4 Impersonation Threats

One reason for unsuccessful e-Learning is the lack of completely trustable, secured, protected and cheating-free e-Examination [3, 21, 24–28]. Many studies report that cheating is common in education [30–32]. Others report that around 70% of American high-school students conduct cheating in at least one exam, where 95% are never caught [33]. In addition, twelve studies report an average of 75% of

Fig. 3 Presence-Identity-
Authentication (P-I-A)
goals [25]

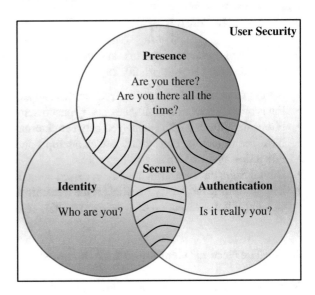

college students cheat [12, 30, 32]. The situation is worse in online exams, where 73.6% of examinees say that cheating is easier and can never be detected [25]. Impersonation is one of cheating-actions that should be prevented or at least detected in e-Examination systems.

Impersonation threats in e-Examination are categorized into three types [25]; Type A, Type B and Type C. Unfortunately, these types alone cannot assure cheating-free e-Exam sessions. Therefore, we proposed a Type D impersonation threat. These types are defined as follows:

1. *Type A* (Connived impersonation threat) supposes that a proctor is necessary. Impersonation might occur in two cases: the proctor could not detect it, or he allowed impersonation by force, sympathy, or for monetary purposes.
2. *Type B* occurs when a student passes his security information to a fraudulent who answers the exam. Username-Password pairs, fall in this type. However, strength of authentication method and existence of a proctor reduce this threat.
3. *Type C* occurs when the real student just login, letting a fraudulent to continue the exam on his behalf. Non-shareable attributes using biometrics approaches, such as fingerprint authentication fall in this greater security-challenging threat.
4. *Type D* might occur such that the real examinee is taking the exam, but another person assists him for correct answers.

2.2   *Authentication Methods in e-Examination*

Several authentication methods are proposed in e-Examination and similar web applications. These methods can be classified into three factors [24].

2.2.1 Knowledge Factors

These factors require a user to know something unique (e.g. a password) that others do not know. With a strong password policy, unauthorized parties cannot access users' information.

2.2.2 Ownership Factors

A user should possess some token that others do not have, such as keys or cards. Unauthorized parties cannot access users' information unless they obtain the required tokens.

2.2.3 Inherence Factors

Referred to as biometrics authentication approaches. They are categorized into two main methods [24]:

- Something a user is: This method utilizes image processing and pattern recognition, e.g. fingerprint, voiceprint, face recognition and retinal pattern.
- Something a user does: the most efficient for continuous user authentication in e-Examination, such as handwriting, keystroke dynamics and mouse dynamics.

Although some of the mentioned authentication methods are highly reliable, they have some drawbacks when used in e-Examination, as summarized in Table 1.

Inherence factors (e.g. biometrics authentication approaches) represent the most powerful approaches of authentication, since they are very hard to be fabricated. Therefore, these approaches are used in e-Examination schemes, as will be discussed in Sect. 2.3. We focus on three approaches, since they are used in our e-Examination model.

Table 1 Drawbacks of user authentication methods in e-Assessment (Extracted from [24])

Authentication method	Drawbacks
Knowledge factors	1. If the password is given away, the security policy will be cancelled 2. A password is requested once at login. They are never trusted for continuous authentication
Ownership factors	1. If the token is passed to others, the scheme is circumvented 2. A token is requested once at login. They cannot be trusted for continuous authentication
Inherence factors	1. They are more reliable, but require special hardware 2. They are unreasonably intrusive, expensive and difficult to implement 3. Some approaches repeat authentication continuously, but they are not fully trusted 4. They are never trusted in case of getting assistance from others

2.2.4 Fingerprint Authentication (FPA)

Fingerprint authentication (FPA) is implemented in many web applications. For instance, it is implemented for user authentication in e-Examination [12, 24, 34]. Nowadays, fingerprint is commonly used for user login, and manufacturers produced a fingerprint mouse, such that a finger scanner is compacted under the thumb for continuous authentication. In addition, reliable fingerprint servers are available with false reject rate (FRR) of 0.01% [5]. The main steps of fingerprint biometrics authentication proceed as follows [34]:

- Creating user-ID, scanning each user's thumb and storing it in a secure server.
- Log in using user-ID and fingerprint via a scanner device when prompted.
- The device will be disabled and the user will be able to access sensitive data.

Two metrics of security level are defined for fingerprint, as shown in Eqs. (1) and (2) [35]:

$$FAR = \frac{IFA}{TNIT} \tag{1}$$

$$FRR = \frac{CFR}{TNCT} \tag{2}$$

Where, FAR is the false acceptance rate, IFA is the ratio of impostors that were falsely accepted, $TNIT$ is the total number of tested impostors, FRR is the false rejection rate, CFR is the ratio of clients that are falsely rejected, and $TNCT$ is the total number of tested clients. FAR measures the probability that an impostor is falsely accepted, whereas FRR measures the probability that a valid user is rejected.

2.2.5 Keystroke Dynamics Authentication (KDA)

KDA proposes that typing rhythm is different from a user to another. It was proposed with five metrics of user identity verification [24]:

- Typing speed, measured in characters per minute.
- Flight-time between two keys up, including the time a user holds on a key.
- Keystroke seek-time that is required to seek for a key before pressing it.
- Characteristic sequences of keystrokes, i.e. frequently typed sequences of keys.
- Characteristic errors, i.e. the common errors made by a user to be identified.

Correlation is used to measure similarity among the features of the saved templates and the stroked keys, as shown in Eq. (3) [24].

$$r = \sum_{i=1}^{n} (k_i * t_i) \Big/ \sqrt{\sum_{i=1}^{n} k_i^2 * \sum_{i=1}^{n} t_i^2} \tag{3}$$

Where, r is the correlation, k is a vector of length n which stores flight-time of the template, t is a vector of length n which stores flight-time of the captured keys, and $i \in k, t$ is the flight-time between two keystrokes.

2.2.6 Video Matching Algorithm

This algorithm is proposed originally for video search [36–38], but it can be used for continuous authentication and auto-detection of cheating actions. The examinee's video is matched against his stored template using tree-matching, as shown in Fig. 4. A video is divided into a number of scenes in a structured-tree; each consists of groups of relevant shots. The matching process moves level-by-level in a top-down manner, where similarity is calculated using color histogram and shot style. The algorithm uses a maximum order sum function to compute similarity in four steps [36]:

- Initialize a matrix D with zeros for all elements.
- Fill the matrix according to Eq. (4).

$$D (i + 1, j + 1) = \max (D (i, j), childSim (i, j), D (i, j + 1)) \qquad (4)$$

Where, D is the matrix, $max()$ is the maximum function, and $childSim()$ is the child similarity function.

- Locate the sum of child similarity for the optimal match by Eq. (5).

$$sum = D (numRow + 1, numCol + 1) \qquad (5)$$

Where, sum is the sum of child similarity, $numRow$ is the number of rows, and $numCol$ is the number of columns.

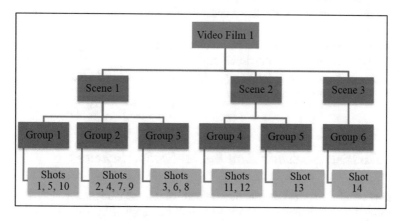

Fig. 4 Structured video tree [36]

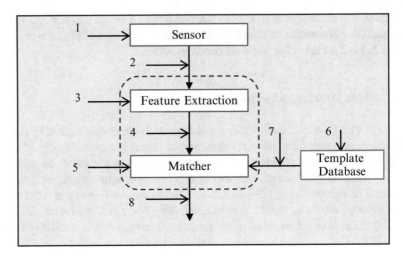

Fig. 5 Attack points in biometrics authentication systems [28]

- Normalize the sum, as shown in Eq. (6).

$$FeatureSimilarity = \left(\frac{sum}{numRow} + \frac{sum}{numCol} \right) /2 \qquad (6)$$

Where, *FeatureSimilarity* is the feature similarity of the current level, *sum* is the sum of child similarity, *numRow* is the number of rows, and *numCol* is the number of columns.

Although biometrics authentication methods are the most efficient ones, still they are vulnerable to cyberattacks. Figure 5 depicts eight points of possible attacks as a result of one or more of the following four risks [28]:

- *Fake input* using either artificial or dummy fingers on the sensor.
- *Low-quality input/imprint* that makes it difficult to extract the minutiae points accurately.
- *Biometric-database modification* including features of fingerprint templates stored on enrollment for the first time.
- *Feature-extractor modification* that might result from attacks leading to incorrect rejection/acceptance of fingerprints.

2.3 *Existing e-Examination Authentication Schemes*

Existing solutions for e-Examination authentication are categorized into five main categories [25]. The same categorization will be adopted in Sect. 4 to conduct a comparison between these schemes and ours.

2.3.1 Proctored-Only Scheme

This approach requires a proctor or a testing administrator to monitor the exam-takers during their examinations. A case study was conducted in which 200 students set for e-Exam in a computer laboratory using WebCT [39]. Three adjacent laboratories were dedicated with three proctors concurrently. A WebCT expert circulated between labs for technical support. Authentication was easy since the proctors were the tutors of the course who knew their students very well. Login ID and password were kept with proctors who distributed them during each session.

Another supporter to this scheme raised three serious problems in e-Assessment [30]; taking assessment answers in advance, unfair retaking of assessments, and unauthorized help. One important caution to reduce the first problem is random selection of the questions for each student out of a large pool, as in Eq. (7) [30].

$$P = \frac{M^2}{N^2} \tag{7}$$

Where, P is the expected overlap between questions in two random exam sets, M is the exam size (i.e. number of questions), and N is the pool size (i.e. number of questions in the pool). In other words, to raise the probability of choosing a distinct set of questions for each student, at least, Eq. (8) should be satisfied [30].

$$N = S * M \tag{8}$$

Where, N is the pool size, S is the number of students to set for the exam, and M is the exam size. Proponents of this scheme consider proctor-based e-Assessment suitable, since it promotes identity and academic honesty [18, 30, 39].

2.3.2 Unimodal Biometrics Scheme

This scheme employs a single biometrics approach for authentication. For example, web authentication, based on face recognition, is used for the verification of student identity with BioTracker that can track students while doing their exams at home. BioTracker can be integrated with LMS, where three concepts are investigated; non-collaborative verification, collaborative verification and biometrics traces [40].

Handwriting approach is another example of this scheme, where a pen tablet is used for writing the most used characters in multiple-choice questions [41]. The written characters are compared with templates that have been taken before the exam [41]. Figure 6 depicts the structure of another similar approach that employs Localized Arc Pattern (LAP) method [41]. It identifies a writer based on one letter written on a piece of paper. It is adapted for multiple-choice e-Exams to recognize an examinee by his handwritten letters [41].

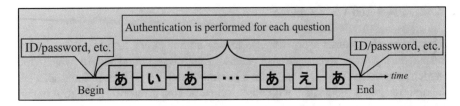

Fig. 6 Handwriting authentication using LAP [41]

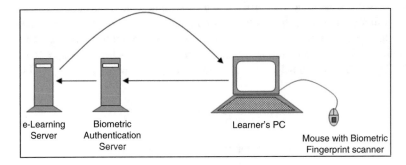

Fig. 7 Fingerprint for e-Exam user's authentication [12]

Another example takes multiple random fingerprints of the examinee throughout the e-Examination period, as shown in Fig. 7. It prevents plagiarism (i.e. pretending to be another examinee) [12, 28].

2.3.3 Bimodal Biometrics Schemes

In order to achieve better security in e-Examination, multiple biometrics approaches can provide reliable user authentication during the exam rather than instantaneous login. For instance, fingerprint is combined with mouse dynamics, where fingerprint is used for login, while mouse dynamics is used for continuous authentication during the exam, as shown in Fig. 8 [42]. Mouse dynamics approach extracts some features from mouse actions that vary from a user to another. This includes motion, click and double-click speed [42].

Fingerprint with head-geometry detection represents another combination, where a webcam captures the pictures of examinees during an e-Exam, and feature extraction and matching with the stored templates are performed [25, 43]. Its structure consists of three modules, as depicted in Fig. 9; (re)authentication ensures the examinee correctness, tracking determines an examinee's position and location, and classifier utilizes the information generated by the tracker to provide risk levels.

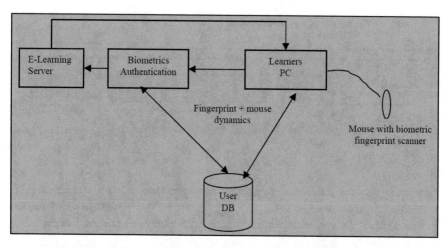

Fig. 8 Bimodal biometrics approach using fingerprint and mouse dynamics [42]

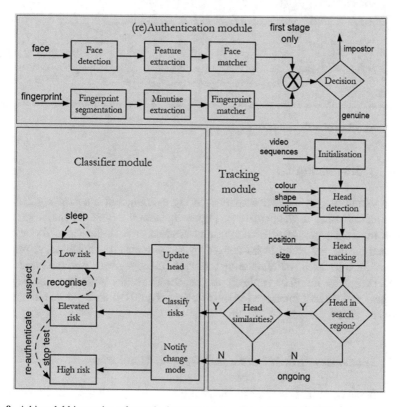

Fig. 9 A bimodal biometrics scheme (e-Assessment security architecture) [25]

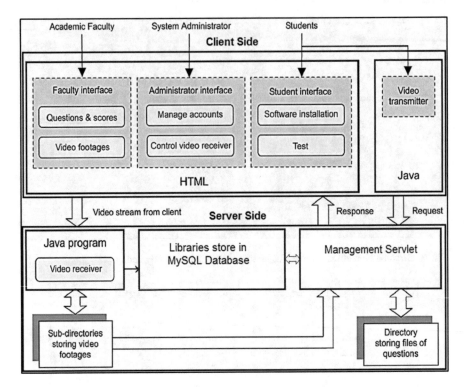

Fig. 10 Structure of e-Test scheme [44] (Adapted)

2.3.4 Video Monitoring

Video monitoring of student activities during examination are also applied in e-Examination using a webcam [44]. Figure 10 shows e-Test scheme's structure, which requires a password for login, and random or periodic video footages are captured during the exam to be revised by the proctor after finishing. It provides three main interfaces: administrator interface (manages user accounts and video receiver), faculty interface (uploads exams, shows results and views the captured footages) and student interface [44]. Unfortunately, this scheme needs extra efforts to watch video footages.

2.3.5 Biometrics Authentication and Webcam Monitoring

This scheme combines fingerprint and real-time video-monitoring, as illustrated in Fig. 11 [35]. When connection to the server is established, the examinee is asked to scan his fingerprint. If it matches the stored one, he can continue. When an exam starts, the webcam streams video to the server to monitor the exam-taker. On mismatch, the exam is interrupted and processed as it is.

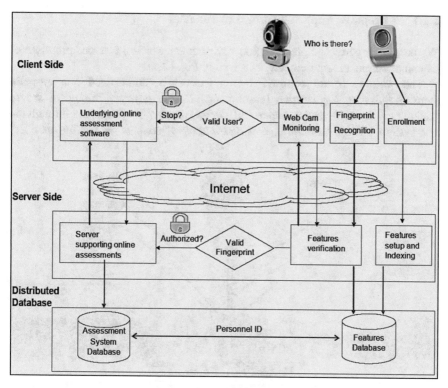

Fig. 11 Structure of combined fingerprint and video-monitoring in e-Examination [35]

3   The Proposed e-Examination Model

Our proposed e-Examination schemes are implemented as modules within a LCMS
(i.e. Moodle). In order to achieve the desired security level, the whole environment
should be secured. For instance, hardware, software, network infrastructure, operating system, e-Learning system, e-Examination system and other valuable assets in
educational institutions should be protected against possible attacks.

This section describes the proposed e-Examination model in three subsections.
The first presents the proposed model's requirements, while the second and the third
provide full description of the proposed model (i.e. ISEEU and SABBAH schemes
respectively and their system-development life-cycle including analysis, design, and
implementation).

3.1   The Proposed-Model's Requirements

In this subsection we provide the main requirements of our proposed e-Examination
model, which includes a proposed implementation environment, hardware and
software requirements, a state diagram and a cheating-action list.

3.1.1 A Proposed Implementation-Environment

We propose a perfect implementation environment which meets our proposed e-examination model requirements, as shown in Fig. 12.

The workflow of the proposed environment can be summarized in seven main steps; *Admission, Registration* (enrollment), *Tuition payment* through a secure banking link to the financial system (FS), *Employment of staff members* through the HR system (HRS), *Scheduling and distribution of courses* on faculty members

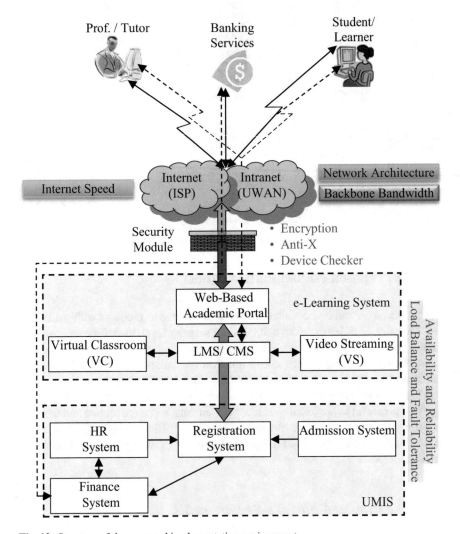

Fig. 12 Structure of the proposed implementation environment

and student access to e-Learning services, *Examination* where e-Exams are auto-corrected and scores are transferred to the registration system, and finally *Salary payment* through a payroll module of the FS and slip delivery.

3.1.2 A Proposed Security Module

The AP and other web applications can be reached via the Internet or the Intranet through a security module shown in Fig. 13. It protects the proposed e-Examination schemes against several vulnerabilities:

1. *Encryption:* Secure Socket Layer/Transport Layer Security (SSL/TLS) protocol is employed to encrypt the data stream between the web-server and the browser.
2. *Firewall/Access Control Lists (ACL):* It blocks unauthorized parties from access to the e-Examination system and prevents any possible connection to assistants.
3. *Cross-Site Scripting (XSS) Detector:* A web vulnerability scanner that indicates vulnerable URLs/scripts and suggests remediation techniques to be fixed easily.
4. *SQL injection Detector:* A web vulnerability scanner that detects SQL injection.
5. *Anti-X (X = Virus, Worm, Spyware, Spam, Malware and bad content):* A reliable firewall/anti-X is installed, and black lists of incoming/outgoing traffic are defined. These lists and definitions of viruses and spyware are kept up-to-date.
6. *Device checker and data collector:* Checks that the authentication devices are functioning properly, and ensures that only one of each is installed, by testing interrupt requests (IRQ) and address space. It collects data about current user and detects violations and exceptions and issues alerts to the e-Examination system.

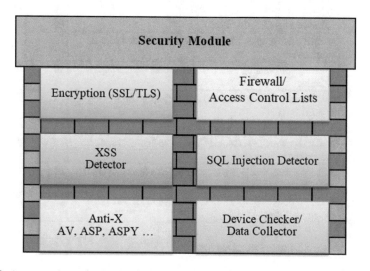

Fig. 13 A proposed security-module

3.1.3 Hardware and Software Requirements

The hardware and software required by the entire integrated e-Learning solution (i.e. the implementation environment) refer specifically to those required to develop the e-Examination scheme. Tables 2 and 3 illustrate hardware and software requirements in both server and client sides

Table 2 Recommended hardware requirements of the proposed schemes

		Proposed scheme	
Side	Hardware requirement	ISEEU	SABBAH
Server-side	1. Media server (MS) with moderate specs	☒	☐
	2. e-Learning server (ELS) with ordinary specifications	☐	☒
	3. Mobile phone module (MPM), e.g. video modem	☒	☐
	4. A video/ fingerprint/ keystroke processing server (VFKPS)	☒	☒
Client-side	1. A personal computer (PC) or a laptop with ordinary specs	☒	☐
	2. 1 Mbps internet connection	☒	☒
	3. A webcam with reasonable resolution	☒	☒
	4. A microphone and a speaker or headphones	☒	☒
	5. A mobile unit (MU) with a camera attached	☐	☐
	6. A biometrics authentication fingerprint-mouse	☐	☒

Table 3 Recommended software requirements for the proposed schemes

		Proposed scheme	
Side	Software requirement	ISEEU	SABBAH
Server-side	1. Windows Server 2008 or RedHat Linux Enterprise RHLE 5.6 or later	☒	☒
	2. Wowza server version 3.0 or later	☒	☒
	3. Moodle server 2.0 or later	☒	☒
	4. PHP 5.3.2 or later, MySQL 5.0.25 or later and Apache web server	☒	☒
	5. The corresponding e-Examination model and security modules	☒	☒
	6. Continuous video matching and fingerprint and keystroke APIs	☐	☒
Client-side	1. Microsoft Windows XP or later	☒	☒
	2. Web browser (e.g. Google Chrome or Internet Explorer 6.0 or later)	☒	☒
	3. Adobe flash media live encoder 3.2 and flash player 10 or later	☒	☒
	4. Continuous fingerprint, keystroke dynamics and video matching APIs	☐	☒

Server Side

Regarding hardware, our proposed e-Examination scheme requires an *e-Learning server (ELS)* and *a media server (MS)* with *sufficient specs* based on the number of expected users. Note that SABBAH scheme requires a *VFKPS server with high storage, memory and processing power.*

Regarding software, a *server operating system,* and *a streaming server* are required. Also, the e-Examination schemes are installed as modules over *Moodle 2.0 server* with suitable *application programming interfaces (APIs).*

Client Side

Hardware requirements for both ISEEU and SABBAH schemes are; *a personal computer* connected to *1 Mbps or more Internet speed,* a *webcam* with reasonable resolution and *headphones.* Also, software requirements include; *an operating system, a web browser, a media encoder* and *a flash player* plugins. Additionally, SABBAH requires *a biometrics mouse* with continuous fingerprint-scanner and its *suitable APIs.*

3.1.4   A Proposed State Diagram

In traditional exams, proctors control and manage exam sessions. They distribute exam papers, announce the start and termination times, monitor and terminate exams or report cheating actions and violations. Similarly, in our proposed schemes, a proctor or the system itself should manage the exam sessions using a state diagram with four possible states (2-bits) with an initial Null state, as shown in Fig. 14.

Transition of the states goes into eight steps throughout an e-Exam for both of our proposed schemes, except that SABBAH replaces *Proctor* with *VFKPS,* as follows:

1. *Null-Ready:* before starting an e-Exam, its state is *Null,* where all records, including number of attempts, are null. When the examinee opens quiz block in Moodle, he is asked to submit his ID, while the "Attempt now" button is dimmed. At this point, the exam state changes to *Ready '00'* and the proctor is notified.
2. *Ready-Start:* if the proctor validates the examinee's identity, he clicks "Accept" in his control toolbox to cause a transition from *Ready* to *Start '01'.* At this point, the "Attempt now" button is enabled, and the examinee is notified to start.
3. *Ready-Null:* if the proctor suspects an examinee is not the correct one, he clicks "Reject", causing a transition from *Ready* to *Null.* So, the examinee should retry.
4. *Start-Pause:* on exceptions, such as webcam removal, the exam is paused for exception handling, and its state is changed from *Start '01'* to *Pause '10'.*

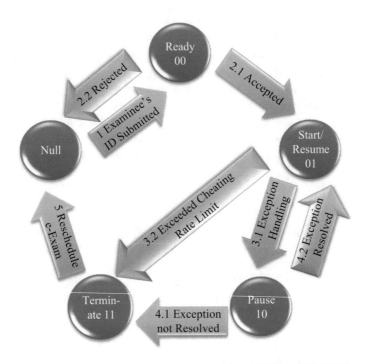

Fig. 14 State diagram of the proposed e-Examination models (ISEEU and SABBAH)

5. *Start-Terminate*: if a violation limit is exceeded, the exam terminates abnormally with a total score of 0, causing a transition from *Start '01'* to *Terminate '11'*. Termination also occurs if an examinee submits his exam or the time is over.

6. *Pause-Resume*: if an exception is handled, the state transits from *Pause '10'* back to *Resume '01'*, and an examinee can continue his e-Exam.

7. *Pause-Terminate:* if exception handling fails for a number of tries, a transition from *Pause '10'* to *Terminate '11'* terminates and reschedules the e-Exam.

8. *Terminate-Null*: if an e-Exam is rescheduled for any reason, its state transits from *Terminate '11'* to *Null*. This deletes all the records of that examinee as if he has not yet conducted his exam. He can re-attempt the exam accordingly.

3.1.5   A Proposed Cheating Action List

In our schemes, we propose a list of cheating actions to measure violation rate and decide penalties, such as deduction from the total score, as illustrated in Table 4.

We conducted a survey with a sample of 50 experts of security, proctoring and education. Each expert was asked to assign a weight of 1–5 to each violation, where violations with higher risks are assigned higher weights. Then, the weights are averaged, approximated and normalized to obtain the final weights of risks.

Table 4 The most popular cheating actions with average risks

No.	Violation/cheating action	Average risk (des. order)
1	Someone else replaced him	0.40
2	Someone else (assistant) is sitting beside him	0.35
3	Redirecting the webcam or disabling it	0.30
4	Incoming/Outgoing calls (Mobile or Telephone)	0.25
5	Send/Receive messages (SMS, MMS, etc.)	0.20
6	Using PDAs (Calculator, iPhone, Android, etc.)	0.20
7	Looking at a textbook or a cheat sheet	0.20
8	Talking with someone	0.10
9	Looking around	0.10
10	Hiding face with hand or another object, or by sleeping on desk	0.10

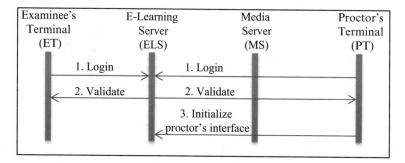

Fig. 15 Pre-operation of ISEEU

3.2   Interactive and Secure e-Examination Unit (ISEEU)

Interactive and Secure e-Examination Unit (ISEEU) is implemented on Moodle
using PHP, MySQL, HTML, AJAX, and other scripting languages. A media server
is also used for streaming exam sessions to the corresponding proctor and for
recording. ISEEU provides online e-Exams with new features such as interaction
between proctors and examinees, and minimizing cheating including access to
resources and impersonation threats. ISEEU is one of the simplest and the most
efficient approaches of e-Exam authentication. An examinee himself, rather than
one or part of his organs, is identified continuously throughout his e-Exam. Security
and reliability wise, it can be more efficient than in-classroom sessions.

Pre-Operation of ISEEU starts with a few steps that prepare for e-Examination,
as shown in Fig. 15. The main steps of pre-operation are:

1. *Login*: Clients (e.g. examinees or proctors) start with login to the e-Learning
   portal (e.g. Moodle), using username and password.
2. *Validate examinee/proctor*: if an examinee's or a proctor's identity is initially
   validated by the e-Learning server (ELS), he is granted access to the system.

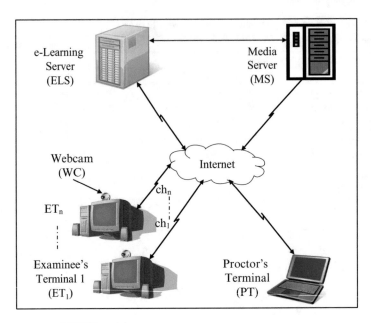

Fig. 16 Structure of ISEEU model

3. *Initialize proctor's interface*: On the main interface of the e-Course, users with proctor role clicks a link to a monitoring interface, which consists of multiple video screens; one per examinee. It can be zoomed in/out when a violation action is suspected. Also, it contains a control toolbox and a dropdown violation-list.

3.2.1 ISEEU Structure

ISEEU employs a webcam attached to an examinee's terminal (ET) that streams his exam session to a media server (MS) through the Internet, as shown in Fig. 16.

The MS, in turn, forwards exam sessions to proctors through the e-Learning server (ELS). Each examinee's session is streamed through his channel, and all the video streams appear on the proctor's terminal (PT). Both examinee and proctor are connected to the ELS and the MS through a security module that protects them against possible attacks.

3.2.2 ISEEU Procedure

The flowchart of ISEEU is shown in Fig. 17. It describes its operation and the procedure of conducting e-Exams. Moreover, the sequence diagram of Fig. 18 interactively clarifies its operation in 18 steps:

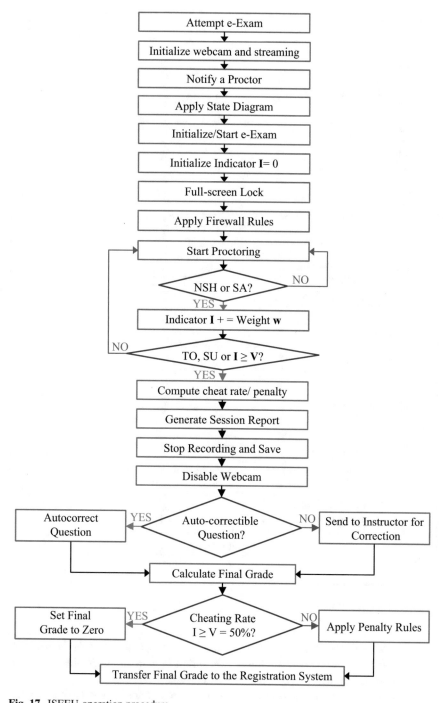

Fig. 17 ISEEU-operation procedure

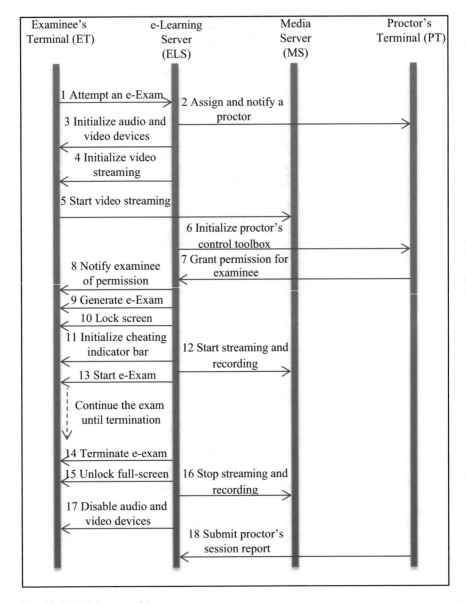

Fig. 18 ISEEU Sequence Diagram

1. *Attempt an e-Exam*: the "Attempt now" button on the exam's interface it disabled. At this point forward, *the state-diagram* controls the transition between different states based on specific events.
2. *Assign and notify a proctor*: the system generates a query to the database on the ELS to determine the corresponding proctor. If he is online, it notifies him.

3. *Initialize audio and video devices*: initialize the examinee's audio and video devices (e.g. headphones and webcam) and asks him to calibrate them.
4. *Initialize video streaming*: both the media encoder on the examinee's terminal (ET) and the media server (MS) are initialized, and their connection is check.
5. *Start video streaming*: a channel is assigned and a connection is established between the ET and the MS. The examinee is asked to calibrate the webcam until his face is top-centered, before his video starts streaming.
6. *Initialize proctor's control toolbox and notify him*: initializes the control toolbox for both examinee and proctor and activates the required functions.
7. *Grant permission to examinee*: the examinee waits permission from his proctor to start. While waiting, an examinee can play a demo of instructions. At this step, all necessary functions are activated. Proctors interact with examinees via chat option in the toolbox and follow a predefined procedure to verify their identity.
8. *Notify examinee of permission*: if an examinee is identified, the proctor approves by clicking "Accept". This enables the "Attempt now" button, and an examinee is notified to start his e-Exam.
9. *Generate e-Exam*: this step randomly generates the questions from a question bank on the ELS that guarantees different questions for each examinee.
10. *Lock the examinee's screen*: immediately, the exam-interface is locked with a full screen, and all functions, that allow access to resources, are disabled.
11. *Initialize cheating indicator bar*: a graphical cheating indicator bar is initialized to 0% ($I = 0$) and appears on the exam's interface.
12. *Start streaming and recording*: the e-Exam session streaming and recording start. It is stored to the MS in order to be revised on uncertainty.
13. *Start e-Exam*: the timer is initialized, the exam session starts and continues until time is over (TO), exam is submitted (SU) or terminated when violation limit ($I \geq V$) is exceeded. On exceptions, such as device failure or disconnection, the device checker handles exceptions before the exam is being resumed.
14. *Terminate e-Exam*: If the exam terminates, it is closed with a relevant warning message. The questions are auto-scored and scores appear to the examinee.
15. *Unlock full-screen*: the examinee's interface is unlocked to its normal state.
16. *Stop streaming and recording*: streaming is stopped and the video is saved.
17. *Disable audio and video devices*: the webcam goes off.
18. *Submit session report*: It generates a session report that contains all violations. A proctor revises it and the recorded session, if necessary, and submits his report.

During an e-Exam, a proctor might pause/resume a session and can generate alerts and violations by choosing from a list. Cheating rate is calculated each time and appears on a cheating indicator bar (I) on the examinee's interface. This rate is accumulated on each issued violation ($I+ = w$), such as no show (NSH) or suspicious actions (SA). It is paused or terminated if a violation rate is exceeded ($I \geq V$).

3.3 Smart Approach for Bimodal Biometrics Authentication in Home-exams (SABBAH)

SABBAH scheme resolves the major challenges of ISEEU, especially that it does not require manual intrusion. The following subsections introduce the new features.

3.3.1 SABBAH Structure and Features

SABBAH scheme comes as an upgrade of ISEEU. We add a bimodal biometrics scheme, which consists of continuous fingerprint and keystroke dynamics. The first employs a mouse with a built-in fingerprint scanner, while the latter employs the keyboard, as depicted in Fig. 19. Another important difference is automation, where the PT is substituted by a Video/FPA/KDA Processing Server (VFKPS).

The new features of SABBAH over ISEEU can be summarized as follows:

- Fingerprint is used for login and for continuous verification. It guarantees the examinee's presence if the webcam fails, and continues while it is being fixed.
- Keystroke dynamics ensure that the actual examinee is typing in essay questions.

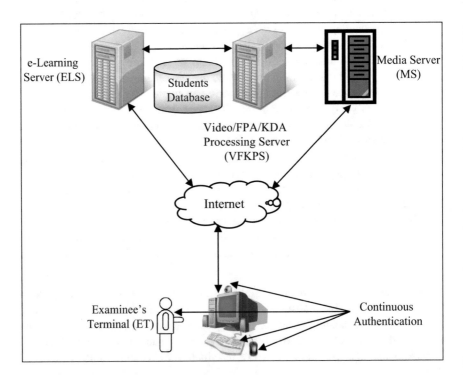

Fig. 19 Structure of SABBAH e-Examination Scheme

- The device checker ensures that authentication devices are the only functional ones, by investigating ports, interrupt requests (IRQ) and memory addresses.
- Firewall/access lists module rejects unknown in/out protocols by closing all ports except the required ones. This prevent communication with possible assistants.
- The VFKPS server automatically substitutes proctors using video comparison.

3.3.2 SABBAH Procedure

SABBAH operates in three phases; enrollment, exam session, and finalization. Again, Fig. 18 describes SABBAH sequence. Most steps are the same except authentication methods and the VFKPS, which replaces the proctor's terminal (PT).

Phase I: Enrollment

This phase starts when a student enrolls e-Learning courses, as shown in Fig. 20:

1. Student's fingerprint is scanned at the registrar's desk.
2. A still photo and a short video are captured by a high-resolution camera.
3. A training set of keystrokes is captured by typing a passage on a dedicated PC.
4. The VFKPS performs feature extraction and saves that on the ELS.

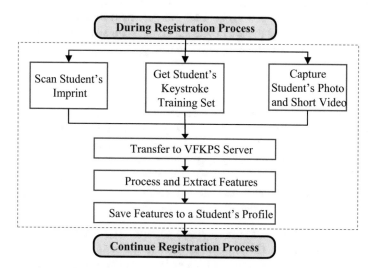

Fig. 20 Phase I (Enrollment)

Phase II: e-Examination Session

In this phase, an exam passes into four possible states in the VFKPS as illustrated in Fig. 21. When an examinee opens his video interface, he is asked to check video and audio settings. Then he submits his ID, while the "Attempt now" is dimmed.

Initialization

This phase enables, checks, configures calibrates devices, and login to the ELS and the e-Examination system takes place. It consists of three steps; FPA, KDA and video initialization, as shown in Fig. 21.

FPA Initialization When a student opens the login screen, the fingerprint scanner is enabled, and he is asked to enter his imprint on the on-mouse scanner. The VFKPS runs a matching algorithm with the saved imprint. If login succeeds, the examinee moves to the next step. Otherwise, he should just retry.

KDA Initialization This starts if the FPA succeeds, where a multimedia demo appears with instructions, and the examinee is asked to type a short paragraph. Keystrokes are transferred to the VFKPS for feature extraction and matching with the stored ones. If they match, he moves to the next step, otherwise, he should retry.

Video Initialization After KDA initialization completes, the examinee is moved to a blank video screen. Then, the webcam is enabled and his video appears, and the examinee is asked to calibrate video and audio devices. He is reminded of chat service with technical-support agents 24/7 online. The matching algorithm extracts the features and compares them with the stored ones. Finally, if matched, he moves to the next step, otherwise, he will just keep trying.

Operation

On exam initialization, the system chooses randomly from a large pool of questions to minimize the chance for examinees to get similar questions. The pool size should be at least equal to the number of examinees times the exam size. In this phase, the exam actually starts, and so the timer's countdown. Also, a full screen locks the examinee's desktop and the security module closes the ports to prevent access to related resources from local disks, internet, remote desktops or remote assistants. The cheating indicator is initialized to zero ($I = 0$).

FPA Operation The fingerprint scanner captures the imprint in two modes; randomly or periodically. The imprints are transferred to the VFKPS for continuous matching. If matched, a new imprint is captured. Otherwise, it continues trying and moves to KDA matching.

KDA Operation The examinee's activities on the keyboard are continuously captured and sent to the VFKPS for matching. If matched, a new keystroke set is captured. Otherwise, it continues trying and moves to video matching.

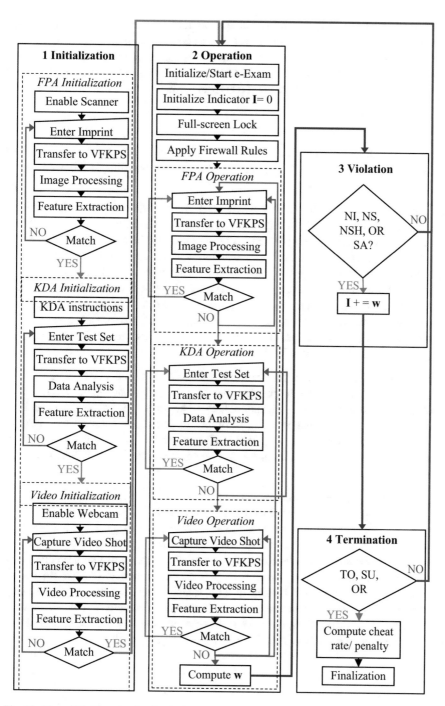

Fig. 21 Phase II (e-Examination Session)

Video Operation This step takes video shots randomly or periodically and sends them to the VFKPS for feature extraction and continuous matching. If matched, a new video shot is captured and the operation is repeated. Otherwise, it continues trying and repeats the operation cycle from the beginning. It also moves to the next step to add a violation with a weight (w) from the list.

Violation

It occurs when some rules are violated, either by the system or the examinee. The exam saves status, pauses, if necessary, and exception handling is followed before the exam can be resumed.

System's Violation This occurs when the keyboard, the mouse or the webcam stops responding, turned off, removed or duplicated. The device checker can detect such violations. Also, power off, internet disconnection and application or operating system errors are considered. If the examinee did not cause any, these errors will not be treated as cheating actions, and no penalty will be applied. Following the state diagram, the exam saves its state, pauses and notifies an agent in the technical support unit (TSU). When the violation reason is detected and corrected, the exam resumes. Restart and shutdown of either hardware or software are allowed to fix the problem. For security issues, the examinee cannot review the previously answered questions after the system is back operational.

Examinee's Violation This occurs when the examinee violates instructions, cheats or tries to cheat. Violation might be impersonation, getting assistance from others, or access to exam resources or material, etc. The violations are weighted and the cheating rate is accumulated using weights (w) and represented on the cheating indicator (I). A violation is weighted if all devices fail to recognize a user, or if a suspicious action (SA) is detected. The examinee's violations in SABBAH can be:

- *FPA violations* include unmatched fingerprint with the stored one, and no imprint (NI) is captured for a predefined period in questions that require a mouse.
- *KDA violations* include unmatched features of keystrokes and keystroke absence, i.e. No strokes (NS), for a predefined period in essay questions.
- *Video violations* are not limited to face and/or head are unmatched, specific parts of the body do not show (NSH) for a predetermined number of tries, suspicious actions (SA) and moves (looking around, sleeping, bending, etc.), and producing noise, speech or voice. On each violation, the system generates relevant warning messages, and the cheating indicator increments ($I+ = w$). If the resultant rate exceeds some violation limit ($I \geq V$), it moves to termination.

Termination

In this phase, the examination session actually terminates, as follows:

Normal Termination This occurs either when the exam's time is over (TO), or when an examinee submits all questions (SU). In both cases, the system saves the session's status-report and video recording. The report includes all violations and the total cheating rate. Finally, it moves to finalization phase that unlocks the full screen, turns off authentication devices and applies penalties.

Abnormal (Cheating) Termination Each time the examinee commits a violation, it appears on his cheating indicator bar. When this rate exceeds a specific limit, say $(I \geq 50\%)$, the exam automatically terminates with a zero grade. In fact, this rate depends on the institution's rules and can be configured in the system settings. After termination, the same procedure in (1) is followed.

Phase III: Finalization

The flowchart of this phase is shown in Fig. 22. It includes grading, applying penalties, reporting, and transfer of scores.

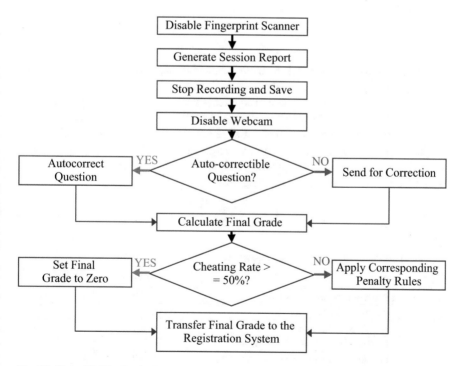

Fig. 22 Phase III (Finalization)

After an exam terminates, a session report is generated, and session recording is stopped and saved. Then the exam is corrected, where auto-correctable questions (e.g. matching or multiple-choice) are corrected. Otherwise, they are sent to the instructor for manual correction. The total grade is recalculated for all questions.

Based on the cheating rate in the generated report, a penalty is applied. For instance, a penalty of 50% cheating rate or more can be a total of zero. Actually, this can be set as a preference according to each institution's policy. In case of uncertainty, a monitoring specialist is notified and given access to an examinee's relevant data and the recorded session. He revises them, evaluates cheating rate, and submits his report, which is considered in recalculating the total grade. The final grade is then transferred to the registration system automatically.

4   Results and Conclusion

In this section, we provide results of a comprehensive risk-analysis that we have conducted against seven security risks. The objective this analysis is to measure our proposed schemes' security and their ability for cheating prevention and detection compared to the previous schemes. The section is divided into five subsections; results, discussion and evaluation, conclusion, challenges and future work.

4.1   Results

For more accuracy in the risk analysis, each risk measures several threats. The distribution of priority on security risks is illustrated in Table 5. It is shown that *Preventing access to resources* achieved the highest priority with a weight of *0.708*, whereas *e-Learning environment security* achieved the lowest priority with a weight of *0.533*. Note that the 5th and the 6th risks have the same priority of *0.567*.

The average score of each scheme is weighted by Eq. (9) [45, 46], where each score is multiplied by its corresponding weight and summed. Then, their summation is divided by the summation of all weights.

$$T_s = \frac{\sum_{i=0}^{n}(W_i S_i)}{\sum_{i=0}^{n} W_i} \tag{9}$$

Table 5 Distribution of priority on security risks

No.	Risk	Weight	Priority
1	Preventing access to resources	0.708	1
2	Satisfying C-I-A goals	0.654	2
3	Satisfying P-I-A goals	0.631	3
4	Impersonation threats (C, D, B, A)	0.613	4
5	Interaction with examinees and feedback	0.567	5
6	Redundancy/fault tolerance of auth. methods	0.567	5
7	Security of e-Learning environment	0.533	6

Where, T_S is the total weighted-score of a scheme s, W_i is the priority weight of a risk i, S_i is the score of a scheme for a risk i, and n is the total number of risks.

Final results of this risk analysis are shown in Tables 6 and 7. They are also represented by bar charts in Figs. 23 and 24 respectively. In order to compare security of the entire e-Examination schemes, the average scores have been computed. For briefing and simplicity, the scheme with the best score in each category is considered to represent it. Also, both of our proposed schemes are maintained. The remaining lower-score schemes in each category are neglected. Then, results of the remaining schemes are listed within their categories in Table 7. Moreover, their bar charts are illustrated in Fig. 24.

The Italicized scores in the tables indicate that a scheme failed to resolve a risk, i.e. its score is less than 50%. The last row in Tables 6 and 7 show the weighted-average scores of each scheme or category based on Eq. (9). This measures the impact of the risks' priorities shown in Table 5.

4.2 Discussion and Evaluation

Before discussion and evaluation, we start with a previous evaluation study to be compared with our comprehensive evaluation. Apampa [25] has conducted the previous evaluation of the existing schemes against the impersonation threats (e.g. Types A, B and C), as depicted in Table 8. "Yes" means that a scheme solves a threat, "No" means that it is susceptible to a threat, and "SP" stands for strong potential [25].

The previous evaluation show that [25]:

1. The first scheme is vulnerable to Type A threats. Another person else can conduct an e-Exam with connivance of a proctor.
2. The second solves Type B and prevents scenarios of pretending to be the real examinee. It is feasible for Type C if continuous authentication is performed.
3. In the third, fingerprint with mouse dynamics solve Type B, but unclearly solve Type C due to delay in mouse data-capturing. Alternatively, fingerprint with face-geometry detection minimize Type B and Type C threats.
4. The fourth scheme is vulnerable to Type A, B and C. Live video monitoring will fail if the proctor is unfocussed, and video revision needs extra efforts.
5. Fingerprint of the fifth scheme solves Type B, while video monitoring is unclear. Moreover, security will be broken if the webcam is moved.

Unfortunately, the previous evaluation is not accurate, since it considers that any scheme with human interaction is susceptible to connived Type A, and a successful attack creates a route for a Type C impersonation. However, it is impossible to prevent human interaction. For instance, developers, database administrators and network administrators have full access to the e-Examination system, and if connived, the system is circumvented. Moreover, our proposed Type D impersonation was not taken into account, and other important security issues such as access to

Table 6 ISEEU and SABBAH compared with the previous e-Examination *schemes* against security risks

No.	Security risk	Previous scheme								Proposed scheme	
		Proc-LAB	BWA-FR	eTest-LAP	Theo-CFP	FP-MD	FP-HGD	SIE-VM	FP-VM	ISEEU	SABBAH
1	Preventing access to resources	54.5%	15.2%	15.2%	24.2%	24.2%	24.2%	63.6%	63.6%	100%	100%
2	Satisfying C-I-A goals	66.7%	66.7%	66.7%	66.7%	66.7%	66.7%	66.7%	66.7%	66.7%	100%
3	Satisfying P-I-A goals	100%	33.3%	66.7%	66.7%	66.7%	66.7%	66.7%	66.7%	66.7%	100%
4	Impersonation threats	100%	35.7%	71.4%	71.4%	71.4%	71.4%	71.4%	100%	100%	100%
5	Interaction and feedback	100%	30.8%	30.8%	30.8%	30.8%	30.8%	30.8%	30.8%	69.2%	100%
6	R/FT of auth. methods	100%	8.7%	26.1%	30.4%	47.8%	47.8%	21.7%	43.5%	43.5%	73.9%
7	Environment security	100%	20.0%	20.0%	20.0%	20.0%	20.0%	20.0%	20.0%	100%	100%
	Average score	88.7%	30.0%	42.4%	44.3%	46.8%	46.8%	48.7%	55.9%	78.0%	96.3%
	Weighted-average score	87.4%	30.5%	42.8%	44.9%	47.2%	47.2%	50.3%	57.3%	78.4%	96.5%

Table 7 Comparison of our proposed schemes (i.e., ISEEU and SABBAH) and the previous *categories* (i.e., schemes are combined into their categories) against security risks

No.	Security risk	Previous-scheme category						Proposed scheme	
		Proctored-only	Unimodal biometrics	Bimodal biometrics	Video monitoring	Biometrics with VM	ISEEU	SABBAH
1	Preventing access to resources	54.5%	18.2%	24.2%	63.6%	63.6%	100%	100%
2	Satisfying C-1-A goals	66.7%	66.7%	66.7%	66.7%	66.7%	66.7%	100%
3	Satisfying P-I-A goals	100%	55.6%	66.7%	66.7%	66.7%	66.7%	100%
4	Impersonation threats	100%	59.5%	71.4%	71.4%	100%	100%	100%
5	Interaction and feedback	100%	30.8%	30.8%	30.8%	30.8%	69.2%	100%
6	R/FT of auth. methods	100%	21.7%	47.8%	21.7%	43.5%	43.5%	73.9%
7	Environment security	100%	20.0%	20.0%	20.0%	20.0%	100%	100%
	Average score	88.7%	44.3%	46.8%	48.7%	55.9%	78.0%	96.3%
	Weighted-average score	87.4%	44.9%	47.2%	50.3%	57.3%	78.4%	96.5%

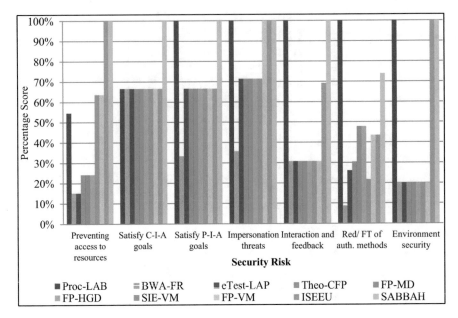

Fig. 23 ISEEU and SABBAH compared with the previous e-Examination *Schemes* against security risks (Percentage Scores)

resources, C-I-A and P-I-A goals were not considered. Therefore, we present a more accurate and comprehensive evaluation below.

Our evaluation and discussion assume that the proctor-based scheme (i.e. traditional) has the highest security in the optimal case. In general, e-Examination schemes compete with it to achieve, at most, the same security. The proctored-only (Proc-LAB) scheme is the most similar to a traditional one, except that it is computer-based and conducted in computer labs.

Nevertheless, *SABBAH scheme* (i.e. our proposed scheme) achieved the *first rank* and the *best security* with an average score of *96.3%*. The following justify this distinct rank of SABBAH in the measured seven-risks:

1. *Access to exam resources (100%):* Exam resources are protected by several methods:

   - The full-screen lock prevents access to local disks, storage media, Internet, Intranet, LANs and WANs.
   - Video matching detects cheating actions such as copy from textbooks, cheat sheets and PDAs (e.g. Phone Calls/MMS/SMS).
   - Exam questions are randomly generated from a large pool (e.g. a question bank).
   - The exam repository is protected with fingerprint and the security module (e.g. SSL/TLS, firewall, anti-x, XXS/SQLI detection, etc.).

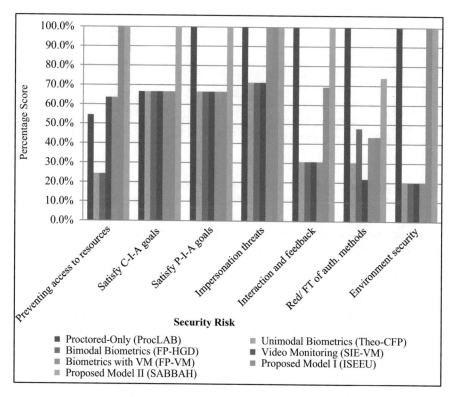

Fig. 24 Comparison of our proposed schemes (i.e. ISEEU and SABBAH) and the previous *Categories* (i.e. combined schemes) against security risks (Percentage Scores)

2. *C-I-A goals (100%)*: Strong authentication and security module assure confidentiality, integrity and availability. Availability might be slightly degraded when the number of examinees exceeds some limit, if not scheduled carefully.
3. *P-I-A goals (100%)*: Continuous authentication using fingerprint, keystroke dynamics and video matching ensure presence, identification and authentication.
4. *Impersonation threats (100%)*: All types of impersonation threats are solved:

   – *Type A*: never occurs, since there is no manual intervention. Also, sessions are recorded for more processing on uncertainty.
   – *Type B*: never occurs, since biometrics authentication is used, such that no way to pass security information to fraudulent persons.
   – *Type C*: similar to type B, continuous authentication is used to prevent allowing fraudulent to get exam-control, since this scenario will be detected.
   – *Type D*: video matching catches local assistants, while remote assistants are prevented by the security module, which closes all vulnerable ports.

Table 8 Impersonation threats and solutions of existing schemes [25]

No.	Scheme category	Solution to impersonation threat		
		Type A	Type B	Type C
1	Proctored-only environment [18, 30, 39]	No	No	No
2	Unimodal biometrics [12, 24, 28, 40, 41]	Yes	Yes	No
3	Bimodal biometrics [25, 42, 43]	Yes	Yes	SP
4	Video monitoring (+password) [44]	No	No	No
5	Biometrics + webcam monitoring [35]	Yes	Yes	No

5. *Interaction and feedback (100%)*: Two effective tools of interaction with exam-inees are provided; the cheating indicator and warning messages. The indicator measures the cheating rate and appears on the exam's interface. At the same time, warning messages appear to examinees on each violation.
6. *Redundancy/fault tolerance of authentication devices (73.9%)*: If one of the three authentication devices fails, another continues working. The device checker, in the security module, checks devices, detects failures and fixes them. It also assures a single device only for each authentication method.
7. *Environment security (100%)*: The security module provides a typical secure environment for the whole system. The LCMS (i.e. Moodle) is protected with SSL/TLS encryption. The firewall, the access lists, the XSS and SQLI detectors reduce attacks. Moreover, attachments are checked for viruses, malware, spy-ware, etc. If any is detected, it will be recovered, quarantined or deleted.

Proctored-only (Proc-LAB) comes next by achieving the second rank with an average score of 88.7%. It scored 54.5 and 66.7% in the 1st and the 2nd risks respectively and 100% in the last five risks. However, proctored-only cannot be considered a pure e-Examination scheme.

Our proposed scheme *(ISEEU) is ranked the third* by achieving an average score of *78%*. Although it failed in the 6th risk with a score of 43.5%, this risk has a lower priority and its impact is not considerable. Justification for this reasonable result is:

1. *Access to exam resources (100%)*: The same as SABBAH, but video monitoring replaces video matching for cheating prevention.
2. *C-I-A goals (66.7%)*: The security module provides high confidentiality and integrity. Availability is degraded when the number of examinees exceeds some limit, since each session needs a new channel. This reserves more memory, CPU and bandwidth. Therefore, it should be scalable and kept under monitoring.
3. *P-I-A goals (66.7%)*: Continuous video monitoring guarantees presence, identifi-cation and authentication, but using username and password to login is still weak and vulnerable.
4. *Impersonation threats (100%)*: All types of impersonation threats are solved:

   – *Type A*: exam sessions are monitored and recorded for uncertainty cases.

Type B: continuous monitoring detects the scenario of plagiarism (i.e. conducting exams on behalf of others) using their security information.x*Type C*: allowing others to get exam-control can be detected with continuous video monitoring.

- *Type D*: video monitoring catches local assistants, and closing vulnerable ports and protocols, other than those required, prevents remote assistants.

5. *Interaction and feedback (69.2%)*: ISEEU employs a cheating indicator bar and warning messages. It also uses text and audio chat for more interaction.
6. *Redundancy and fault tolerance of authentication devices (43.5%)*: It uses passwords for login to the LCMS. Moreover, video monitoring and recording are used for continuous authentication. If the webcam fails, the exam pauses until being fixed. Otherwise it is terminated.
7. *Environment security (100%)*: ISEEU exists in the same environment of SAB-BAH model described above.

Finally, biometrics with video monitoring (FP-VM) ranked the fourth with *55.9%*. Video monitoring (SIE-VM), bimodal biometrics (FP-HGD) and unimodal biometrics (Theo-CFP) failed with *48.7%*, *46.8%* and *44.3%* respectively.

The average scores were used in the previous discussion, in which risk priority was not considered. The weighted-average scores of the seven risks are illustrated in Fig. 25 in descending order. Accordingly, *SABBAH* achieved the *1st rank* with a weighted-average score of *96.5%*, whereas *ISEEU* achieved the *3rd rank* with *78.4%*. Proctored-only achieved the *2nd rank* with *87.4%*. It is being emphasized here that Proctored-only is not a pure e-Examination scheme, since exams could not be conducted at home. The *4th* and the *5th* ranks are achieved by biometrics with VM and video monitoring with *57.3%* and *50.3%* respectively. Finally, bimodal biometrics and unimodal biometrics failed with *47.2%* and *44.9%* respectively.

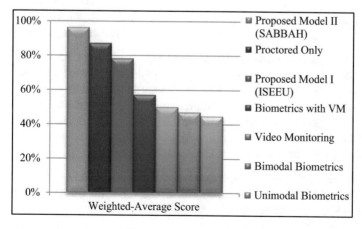

Fig. 25 Weighted-average scores of ISEEU and SABBAH against the previous categories

4.3   Conclusion

It is shown that security is vital for e-Learning systems, since they are operational and contain sensitive information and operations. One of these operations is e-Examination, which has been given higher attention recently. If an institution decides to offer pure e-Courses, it will be faced with untrusted e-Examination. Several efforts have been made in order to provide trustful e-Examination by proposing new schemes to improve security and minimize cheating. Even though, those schemes could not compete with traditional proctor-based examination.

This work contributes to solving the problem of cheating and other security issues, taking into account most of the possible vulnerabilities. It is believed that our proposed e-Examination schemes, i.e. ISEEU and SABBAH, present a major contribution in the field. They are highly secure, simple and easy to be implemented. They are the strongest competitors to the traditional examination scheme, or even beat it with its new features. This is exactly what we proved in our results.

Results show that our proposed schemes proved reasonable security compared with the previous ones. *SABBAH ranks the first*, and *ISEEU ranks the third* in the risk analysis. Results show that the proctored-only scheme (i.e. traditional examination) ranked the second after SABBAH. This rank is considered satisfactory, since our objective was to propose a scheme that competes with the traditional one. The reason for superiority of SABBAH is its higher security and its high ability to prevent/detect cheating actions.

However, the claim that traditional exams are 100% secure is theoretical, and the literature shows that over than 70% of high-school students in USA admit cheating, where 95% of them are never detected. Also, they are inflexible in place and time and economically infeasible. Moreover, all procedures, such as exam delivery, monitoring, grading, scoring and data entry, are manual, which are time-consuming and need manpower.

Both *SABBAH and ISEEU* schemes satisfy the *C-I-A* (Confidentiality, Integrity and Availability) and the *P-I-A* (Presence, Identity and Authentication) goals. Also, they resolve security issues that were neither resolved nor mentioned previously:

- Access to exam resources, such as textbooks, worksheets, local computer, the Internet and remote assistance. Both prevent this by fullscreen locks, interactive-monitoring or video matching, and continuous authentication.
- Our schemes are interactive based on cheating indicator and warning messages to stopp cheating when detected, while the previous schemes lack to this feature.
- Webcam failure pauses the exam until it is fixed to resume, while this scenario leads to system failure in the previous video monitoring schemes.

Regarding plagiarism or impersonation threats, our schemes resolve all types of threats. In addition, they have many advantages over the previous schemes. For instance SABBAH scheme is:

- *Fully automated*: a proctor is no longer needed. Also, grading, cheating penalties, and data transfer all are executed automatically.

- *Fully secure*: solving impersonation threats, satisfying P-I-A goals, interaction with examinees, preventing access to resources, and preventing collaboration.
- *Highly efficient*: a virtual examination session, but with similar efficiency in cheating detection and prevention as actual proctored-sessions.
- *Reliable and redundant*: three synchronized authentication devices. If one fails, another passes. Also, a device might never or rarely been used while using another.
- *Reasonable and relatively inexpensive*: it substitutes proctors who are paid for monitoring.

4.4 Challenges

Although they have many advantages and resolve several security issues, our schemes encounter some challenges. The main challenges can be summarized in:

- *Internet speed/backbone bandwidth and robustness*: high-speed and stable internet connection are required, especially at peak times.
- *Performance and capacity*: they require servers with huge memory and storage/disk space, and SABBAH requires a VFKPS with a high processing power.
- *Implementation complexity*: implementation of automatic video matching is not easy, and its algorithms are still under development. Performance and accuracy in feature extraction and matching are still challenging. Also, users on the client side require a special hardware for fingerprint scanning.
- *Failure penalties*: on unrecoverable failures, the answered questions cannot be reviewed by an examinee after being back, for security issues.

However, who keeps track with the fast advancement in information and communication technology (ICT), discovers that all challenges are to be resolved very soon. For instance, proponents claim that cloud-computing concept fits this type of applications and will overcome many of these challenges. Accordingly, this claim was tested in a small range, where ISEEU was deployed on cloud control; computing platform as a service (PaaS). This test was limited, since we used a trial version with limited services. Regarding special hardware requirement, most of modern computers have fingerprint scanners, and touch-screens can provide more options for secure and interactive e-Examination schemes.

It is being emphasized here that our proposed e-Examination schemes will not provide completely cheating-free e-Exams, but they minimize cheating and impersonation threats. With well-developed functions, we believe that our scheme is more efficient and cost-effective than proctor-based in-classroom sessions.

4.5   Future Work

Our research directions of the future is to improve our e-Examination schemes. In order to achieve this, the following are the main ideas intended to be investigated:

- Implement our proposed schemes in a real environment for some courses. This, actually, aims to measure its acceptance, applicability and security. Accordingly, the schemes should be improved in terms of security and performance. This will enable us to examine their resistance to cyber-attacks such as distributed denial-of-service (DDoS), session highjacking, and man-in-the-middle (MitM), etc.
- Develop an efficient video matching algorithm towards being an important biometrics continuous authentication approach, especially for e-Examinations. It detects examinees' violations and cheating actions automatically.
- Apply all parameters of KDA other than typing speed to improve accuracy, such as flight time, seek time, characteristic errors, and characteristic sequences. This enables us to measure its reliability for continuous authentication.
- Deploy our schemes on the cloud to provide the required computational power, and to evaluate its security and performance in terms of storage, memory and CPU usage.

References

1. E. Kritzinger, "Information Security in an e-Learning Environment", 2006. Last access Dec. 2016. http://sedici.unlp.edu.ar/bitstream/handle/10915/24349/documento_completo.pdf%3Fsequence%3d1
2. Y. Sabbah, "Comprehensive Evaluation of e-Learning at Al-Quds Open University", Internal Report OLC-195/2010, Al-Quds Open University (QOU), Open Learning Center (OLC), 2010.
3. M. Hentea, M. J. Shea and L. Pennington, "A Perspective on Fulfilling the Expectations of Distance Education", Proceedings of the 4th Conference on Information Technology Curriculum (CITC4) Lafayette, Indiana, USA, pp.160–167, October 2003.
4. H. A. El-Ghareeb, "e-Learning and Management Information Systems, Universities Need Both", eLearn Magazine, September 2009. Last access December 2016. http://elearnmag.acm.org/featured.cfm?aid=1621693
5. R. Raitman, L. Ngo and N. Augar, "Security in the Online e-Learning Environment", Proceedings of the 5th IEEE International Conference on Advanced Learning Technologies, Kaohsiung, Taiwan, pp.702-706, July 2005.
6. George M. Piskurich (Ed.), "AMA Handbook of eLearning-Effective Design, Implementation and Technology Solutions", AMACOM American Management Association 2003. Last access December 2016. https://www.questia.com/read/119691314/the-ama-handbook-of-e-learning-effective-design
7. M. Bullen, T. Morgan, K. Belfer and A. Qayyum, "The Net Generation in Higher Education: Rhetoric and Reality", International Journal of Excellence in e-Learning (IJEEL), vol.2, no.1, pp.1-13, February 2009.
8. D. Jonas and B. Burns, "The Transition to Blended e-Learning, Changing the Focus of Educational Delivery in Children's Pain Management", Elsevier, Nurse Education in Practice, vol.[8], no.1, pp.1–7, January 2010.

9. W. H. Rice IV, "Moodle e-Learning Course Development: A Complete Guide to Successful Learning Using Moodle", First edition, Packt Publishing 2006.
10. J. C. Taylor, "Fifth Generation Distance Education", Higher Education Series, Report no.40, The University of Southern Queensland, Global Learning Services, 2001.
11. E. R. Weippl, "Security in e-Learning", First edition, Springer 2005.
12. Y. Levy and M. Ramim, "A Theoretical Approach for Biometrics Authentication of e-Exams", Chais Conference on Instructional Technologies Research, Israel, pp.93-101, 2007. Last access December 2016. http://telem-pub.openu.ac.il/users/chais/2007/morning_1/M1_6.pdf
13. National Institute of Standards and Technology. An Introduction to Computer Security: The NIST Handbook. Special Publication 800-12, October 1995.
14. W. Stallings, "Data and Computer Communications", Eighth edition, Prentice Hall 2007.
15. News, "Popular Sites Push Malware", Computer Fraud and Security, Elsevier, vol. 2010, no. 9, pp. 3, September 2010.
16. News, "The Human Impact of Cybercrime", Computer Fraud and Security, Elsvier, vol. 2010, no. 9, pp. 3-20, September 2010.
17. Official Website of Symantec/Norton, "Norton Cybercrime Report: the Human Impact", August 2016, Last access December 2016. http://us.norton.com/theme.jsp?themeid= cyber-crime_report&inid=br_hho_downloads_home_link_cybercrimereport.
18. E. R. Weippl, "In-Depth Tutorials: Security in e-Learning", eLearn Magazine, vol.2005, no.3, March 2005. Last access December 2016. http://elearnmag.acm.org/featured.cfm?aid= 1070943
19. K. El-Khatib, L. Korba, Y. Xu and G. Yee, "Privacy and Security in e-Learning", International Journal of Distance Education, vol.1, no.4, pp.1-19, 2003.
20. S. Banerjee, "Designing a Secure Model of an e-Learning System- A UML-Based Approach", IEEE Potentials Magazine, vol.29, no.5, pp.[20]-27, September-October 2010.
21. N. H. Mohd Alwi, and I.-S. Fan, "Information Security Threats Analysis for e-Learning", Proceedings of the First International Conference TECH-EDUCATION, Athens, Greece, pp.285-291, May 2010.
22. J. F. Gonzalez, M. C. Rodriguez, M. L. Nistal and L. A. Rifon, "Reverse OAuth: A Solution to Achieve Delegated Authorizations in Single Sign-On e-Learning Systems", Computers and Security, vol.28, no.8, pp.843-856, November 2009.
23. V. Ciriani, S. De Capitani di Vimercati, S. Foresti, S. Jajodia, S. Paraboschi and P. Samarati, "Combining Fragmentation and Encryption to Protect Privacy in Data Storage", ACM Transactions on Information and System Security, vol.13, no.3, pp.1-30, July 2010.
24. E. Flior and K. Kowalski, "Continuous Biometric User Authentication in Online Examina-tions", Proceedings of the 7th International Conference on Information Technology: New Generations, Las Vegas, pp.488-492, April 2010.
25. K. M. Apampa, G. Wills and D. Argles, "User Security Issues in Summative e-Assessment Security", International Journal of Digital Society, vol.1, no.2, June 2010.
26. K. Abouchedid and G. M. Eid, "e-Learning Challenges in the Arab World: Revelations from a Case Study Profile", Quality Assurance in Education, vol.12, no.1, pp.15–27, 2004.
27. A. Marcus, J. Raul, R. Ramirez-Velarde and J. Nolazco-Flores, "Addressing Secure Assess-ments for Internet-Based Distance Learning Still an Irresolvable Issue", Proceedings of the 9th Latin-American Congress of Educational Computing, Caracas, Venezuela, March 2008.
28. S. Alotaibi, "Using Biometrics Authentication via Fingerprint Recognition in e-Exams in e-Learning Environment", In the 4th Saudi International Conference, The University of Manchester, UK, July 2010.
29. Apampa KM, Wills G, Argles D. Towards security goals in summative e-assessment security. InInternet Technology and Secured Transactions, 2009. ICITST 2009. International Confer-ence for 2009 Nov 9 (pp. 1–5). IEEE.
30. N. C. Rowe, "Cheating in Online Student Assessment: Beyond Plagiarism", Online Journal of Distance Learning Administration, vol.7, no.2, summer 2004.

31. A. Lathrop and K. Foss, "Student Cheating and Plagiarism in the Internet Era: A Wake-Up Call", Englewood: Libraries Unlimited 2000. Last access December 2016. http://www.questia.com/PM.qst?a=o&d=111682012

32. M. Dick, J. Sheard, C. Bareiss, J. Carter, D. Joyce, T. Harding and C. Laxer, "Addressing Student Cheating: Definitions and Solutions", ACM Special Interest Group on Computer Science Education Bulletin, vol.[32], no.2, pp.172-184, June 2003.

33. Murdock TB, Miller A, Kohlhardt J. Effects of Classroom Context Variables on High School Students' Judgments of the Acceptability and Likelihood of Cheating. Journal of Educational Psychology. 2004 Dec;96(4):765.

34. A. Kapil and A. Garg, "Secure Web Access Model for Sensitive Data", International Journal of Computer Science and Communication, vol.1, no.1, pp.13-16, January-June 2010.

35. J. A. Hernández, A. O. Ortiz, J. Andaverde and G. Burlak, "Biometrics in Online Assessments: A Study Case in High School Students", Proceedings of the 18th International Conference on Electronics, Communications and Computers, Puebla, Mexico, pp.111-116, March 2008.

36. C. W. Ng, I. King and M. R. Lyu, "Video Comparison Using Tree Matching Algorithms", Proceedings of the International Conference on Imaging Science, Systems and Technology, Las Vegas, USA, pp.184-190, June 2001.

37. X. Chen, K. Jia and Z. Deng, "A Video Retrieval Algorithm Based on Spatio-temporal Feature Curves and Key Frames", Proceedings of the 5th International Conference on International Information Hiding and Multimedia Signal Processing, pp.1078-1081, September 2009.

38. M. S. Ryoo and J. K. Aggarwal, "Spatio-Temporal Relationship Match: Video Structure Comparison for Recognition of Complex Human Activities", Proceedings of the IEEE 12th International Conference on Computer Vision, pp.1593-1600, Kyoto, Japan, September-October 2009.

39. G. Harrison, "Computer-Based Assessment Strategies in the Teaching of Databases at Honours Degree Level 1", In H. Williams and L. MacKinnon (Eds.), BNCOD, vol.3112, pp.257-264, Springer 2004.

40. E. G. Agulla, L. A. Rifon, J. L. Alba Castro and C. G. Mateo, "Is My Student at the Other Side? Applying Biometric Web Authentication to e-Learning Environments", Proceedings of the 8th IEEE International Conference on Advanced Learning Technologies, Santander, Spain, pp.551-553, July 2008.

41. S. Kikuchi, T. Furuta and T. Akakura, "Periodical Examinees Identification in e-Test Systems using the Localized Arc Pattern Method", Proceedings of the Distance Learning and Internet Conference, Tokyo, Japan, pp.213-220, November 2008.

42. S. Asha and C. Chellappan, "Authentication of e-Learners Using Multimodal Biometric Technology", in International Symposium on Biometrics and Security Technologies, Islamabad, Pakistan, pp.1-6, April 2008.

43. Y. Levy and M. Ramim, "Initial Development of a Learners' Ratified Acceptance of Multi-biometrics Intentions Model (RAMIM)", Interdisciplinary Journal of e-Learning and Learning Objects, vol.5, pp.379-397, 2009.

44. C. C. Ko and C. D. Cheng, "Secure Internet Examination System Based on Video Monitoring" Internet Research: Electronic Networking Applications and Policy, vol.14, no.1, pp.48-61, 2004.

45. N-Calculators Official Website, "Weighted Mean Calculator". Last access December 2016. http://ncalculators.com/statistics/weighted-mean-calculator.htm

46. Wikipedia Website, "Mathematical Definition of Weighted Mean". Last access December 2016. http://en.wikipedia.org/wiki/Average

Attribute Noise, Classification Technique, and Classification Accuracy

R. Indika P. Wickramasinghe

Abstract Binary data classification is an integral part in cyber-security, as most of the response variables follow a binary nature. The accuracy of data classification depends on various aspects. Though the data classification technique has a major impact on classification accuracy, the nature of the data also matters lot. One of the main concerns that can hinder the classification accuracy is the availability of noise. Therefore, both choosing the appropriate data classification technique and the identification of noise in the data are equally important. The aim of this study is bidirectional. At first, we aim to study the influence of noise on the accurate data classification. Secondly, we strive to improve the classification accuracy by handling the noise. To this end, we compare several classification techniques and propose a novel noise removal algorithm. Our study is based on the collected data about online credit-card transactions. According to the empirical outcomes, we find that the noise hinders the classification accuracy significantly. In addition, the results indicate that the accuracy of data classification depends on the quality of the data and the used classification technique. Out of the selected classification techniques, Random Forest performs better than its counterparts. Furthermore, experimental evidence suggests that the classification accuracy of noised data can be improved by the appropriate selection of the sizes of training and testing data samples. Our proposed simple noise-removal algorithm shows higher performance and the percentage of noise removal significantly depends on the selected bin size.

1 Introduction

Quality of the data takes an upmost importance in data analytics, irrespective of the type of application. In data classification, high quality data are further important as the accuracy of the classification correlates with the quality of data.

Real-world datasets can contain noise, which is one of the reasons that makes the quality of data low [16, 38]. Shahi et al. [39] consider outliers and noise as the

R. Indika P. Wickramasinghe (✉)
Department of Mathematics, Prairie View A&M University, 100 University Dr, Prairie View, TX 77446, USA
e-mail: iprathnathungalage@pvamu.edu

© Springer International Publishing AG 2017
I. Palomares et al. (eds.), *Data Analytics and Decision Support for Cybersecurity*, Data Analytics, DOI 10.1007/978-3-319-59439-2_7

uncertainty of data. Handling data that are mixed with noise brings a hard time for the analysis. Some organizations allocate millions of dollars per year on detecting errors with data [32]. When the noise is mixed with cyber-security related data, one needs to give a serious attention to handle them due to the sensitive nature of the data.

The attention for cyber-security and protective measures grew faster with the expansion of cyberattacks. Stealing intellectual and financial resources are the main reasons behind the deliberate breaching of computer systems. When the mobile commerce expanded revolutionary, a series of thefts started to creep up in which majority of them are related to credit-cards transactions. In a study, Hwang et al. [19] states that approximately 60% of the American adults avoid doing business online due their concern about misuse of the personal information. Riem [33] reports a precarious incident of shutting down a credit-card site of a British bank called, Halifax due to the exposure of consumers' details. In this regard, measurements to minimize credit-card related frauds are in great demand at present than ever before.

Noise hinders the classification [35, 46, 49] by decreasing the performance accuracy, in terms of time in constructing the classifier, and in the size of the classifier. Thist is why the identification of noise is an integral part in data classification. Noise can be introduced in many ways into online transactions of credit card data. Besides the conventional ways that introduce noise, research indicates that magnetic card chips can be vulnerable at times and can introduce noise to the transaction. Therefore, identification and isolation of noise from the data before analyzing is very important.

Apart from the noise, the shape of the attributes' distributions can make an impact on the quality of data classification. It is a fact that most of the natural-continuous random variables adhere some sort of Gaussian distribution. When the data show a departure from the normality, it is considered as skewed. Osborne [31] points out that mistake in data entry, missing data values, presence of outliers, and nature of the variable itself are some of the reasons for the skewness of the data. Furthermore, Osborne [31] makes use of several data transformation techniques such as square root, natural log, and inverse transformation to convert the non-normal data into normal. There is a strong association between the existence of noise and the skewness of the distribution. It is apparent that skewness of data directly influence outliers. Hence, there should be an important connection between the nature of the skewness of data and the classification accuracy.

Even if someone removes the noise and the skewness of the data, it would not completely reach the maximum data accuracy level in the classification. Selection of the correct classification technique based on the available data, and the use of appropriate sample sizes for training and test data are two of options one can consider for improvement of classification. Though there are several literature findings about the identification of appropriate classification technique, only a handful of findings exists in connection with the selection of suitable sample ratios. Even within the available findings, none of them are related to both cyber-security related data that are mixed with noise. In this chapter we have two broad aims. At first, we study the impact of noise in effective data classification. Secondly, we aim to improve the classification of noisy data. To this end, we consider how skewness, appropriate ratios of the samples, and the classification technique impact on the classification. This study brings the novelty in two ways. According to the author's knowledge, it is rare to find a study aiming to investigate the relationship between above selected features and the classification accuracy. Furthermore, we propose a novel, simple and most importantly an effective noise detection algorithm to improve the classification accuracy. The rest of the chapter is organized as follows: Next Sect. 2 discusses the related work, and Sect. 3 provides the background to classification techniques. In Sect. 4, the dataset is described. The Sect. 5 aims to discuss the issue of attribute noise on classification accuracy. Section 6 investigates how skewness of the data influences the classification accuracy. Section 7 attempts to improve the classification accuracy of the data, which is mixed with noise. This is achieved by using the noise removal and the selection of appropriate sample sizes for the training and testing samples. Then Sect. 8 discusses the results of the study and Sect. 9 concludes the chapter.

2 Related Work

Use of SVM in data classification is not novel and has mixture of opinions regarding the accuracy of it. Sahin and Duman [36] incorporated the knowledge of decision trees and SVM in credit card fraud analysis. Though they found the model based on decision trees outperformed SVM, the difference of performances between both methods became less with the increment of the size of the training datasets. In another study, Wei and Yuan [45] used an optimized SVM model to detect online fraudulent credit card and found their proposed non-linear SVM model performed better than the others. Colas and Brazdil [8] and Fabrice and Villa [13] compared SVM with K-nearest neighbor (kNN) and naive Bayes in text classification. Though they expected that SVM would outperform its counterparts, authors couldn't find SVM as the clear winner. In addition, they pointed out that the performance of kNN continues to improve with the use of suitable preprocessing technique. Scholkopf and Smola [37] and Mennatallah et al. [28] utilized SVM in anomaly detection. In addition to SVM based techniques, other alternative techniques can be found in the literature.

Abu-Nimeh et al. [1] compared six classifiers on phishing data. In their study, authors used Logistic Regression, Classification and Regression Trees, Bayesian Additive Regression Trees, Support Vector Machines, Random Forests, and Neural Networks. According to the outcomes, authors claimed that the performance of Logistic Regression was better than the rest.

As the previous research findings indicate, the nature of the data is imperative for the accuracy of classification. Bragging and boosting are considered as popular classifying trees and random forest was proposed by adding an extra layer to bragging [25]. Díaz-Uriarte and Andres [12] incorporated random forest in gene classification and stated that this technique showed an excellent performance in classification even with the presence of noise in data. Miranda et al. [29] compared three noise elimination algorithms for Bioinformatics datasets and Machine Learning classifiers. In another study, Jayavelu and Bar [20] used Fourier series approach to construct a noise removal algorithm. An effective two-phased algorithm for noise-detection is proposed by Zhu et al. [50]. A novel noise-detection algorithm, which is based on fast search-and-find density peaks is proposed in [48]. The authors clustered the original data before removing the outliers. Another outlier detection algorithm on uncertain data is proposed by Wang et al. [44]. This algorithm is based on the Dynamic Programming Approach (DPA). Cao et al. [6] and Kathiresan and Vasanthi [22] conducted their research about handling noise in credit card data. Lee et al. [24] explored the patterns of credit card transactions in order to classify fraudulent transactions.

When the quality of the data is considered, the shape of the distribution of data is very important as the majority of data analysis techniques assume the symmetric nature of the distribution. Akbani et al. [2] suggested that SVMs do not perform too badly with moderately skewed data, compared to the other available machine learning algorithms. Further studies [27, 34, 40, 41] can be seen in the literature. In addition to the nature of the dataset, the ratio of trainee and test data can impact on the quality of classification. Guyon [14] proposed an algorithm regarding the sizes of training and validation datasets, but the author stated that this framework is not perfect due to the use of simplifying assumptions. Beleites et al. [3] studied about the above issue using learning curves for small sample size situations.

3   Background on Classification Techniques and Measurement Indicators

The aim of this section is to describe the theoretical foundation that is used in this study. First of all, the section starts describing the four types of classification techniques that are used in this study. Secondly, quantitative measurements that are used to compare the four classification techniques are discussed.

3.1   Classification Techniques

Here we consider four types of popular classification techniques, namely Support Vector Machines (SVM), Principal Component Analysis (PCA), Robust Principal Component Analysis (RPCA), and Random Forest. These techniques attempt to categorize the class membership based on the attributes of the given dataset by capturing the hidden patterns of the existing dataset. Ultimately they predict the group membership of novel data.

3.1.1 Support Vector Machines (SVM)

SVM is considered as one of the most popular classification techniques [7], which was introduced by Vapnik [43]. This is based on statistical learning theory and it attempts to separate two types of data (Class A and Class B) using a hyper-plane by maximizing the boundary of the separation, as shown in the Fig. 1.

Consider the n-tuple training dataset, S.

$$S = \{(x_1, y_1), (x_2, y_2), \ldots, (x_i, y_i), \ldots, (x_n, y_n)\}. \tag{1}$$

Here the set of feature vector space is M-dimensional and the class variable is 2-demesioinal. i.e., $x_i \in R^M$ and $y_i \in \{-1, +1\}$. As mentioned before, the ultimate aim of the SVM is to find the optimal hyper-plane that split the dataset into two categories.

If the feature space is completely linearly separable, then the optimal separating hyper-plane can be found by solving the following Linear Programming (LP) problem:

$$Min\ ||w||^2$$

$$s.t. y_i\left((x_i.w) + b\right) \geq 1,$$

$$i = 1, 2, \ldots, n \tag{2}$$

Unfortunately not all the datasets can be linearly separable. Therefore, SVM uses a mapping called the *Kernel*. The purpose of the Kernel, Φ is to project the linearly inseparable feature space into a higher dimension so that it can be linearly separated in the higher space. This feature space transformation is taken place according to the following.

$$w.\Phi(x) + b = 0 \tag{3}$$

Fig. 1 Support vector machine

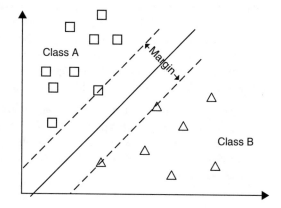

where w represents the weight vector, which is normal to the hyperplane. Following are the most popular existing kernel functions that have been extensively studied in the literature.

- Linear Kernel, $\Phi\left(x_i, x_j\right) = x_i^T x_j$
- Polynomial Kernel, $\Phi\left(x_i, x_j\right) = \left(\gamma x_i^T x_j + 1\right)^d$, for $\gamma > 0$
- Radial Kernel, $\Phi(x_i, x_j) = \exp(-\gamma||x_i - x_j||^2)$, for $\gamma > 0$
- Sigmoid Kernel, $\Phi\left(x_i, x_j\right) = Tanh\left(\alpha x_i^T x_j + r\right)$.

3.1.2 Principal Component Analysis

PCA is an unsupervised and data compression technique used to extract relevant information from a complexed data. The main purpose of PCA is to use in dimension reduction and feature selection [11, 30] by preserving the majority of the characteristics of the initial data, which has been used in various fields. PCA attempts to transform correlated variables into a set of linearly uncorrelated variables in an optimal way. These new set of variables, which is low in dimension is called principal components. This is accomplished via a series of vector transformations as explained in the following algorithm.

Step 1: Begin with the starting variable (the set of features), $X = (X_1, X_2, \ldots, X_n)'$
Step 2: Rotation of the above variables into a new set of variables called, $Y = (Y_1, Y_2, \ldots, Y_n)'$ so that $Y_i's$ are uncorrelated and $\mathrm{var}(Y_1) \geq \mathrm{var}(Y_2) \geq \ldots$ $\mathrm{var}(Y_n) \geq$.
Step 3: Y_i and Y_{i+1} that are constructed so that $\alpha_i' = \left(\alpha_{1j}, \alpha_{2j}, \ldots, \alpha_{pj}\right)$, $Y_i = \alpha_i'X$ and $\sum \alpha_i^2 = 1$.

3.1.3 Robust Principal Component Analysis (RPCA)

PCA is considered as not robust enough to work with data when outliers are present. Halouska and Powers [15] investigated the impact of PCA on noise data related to Nuclear Magnetic Resonance (NMR). They concluded that a very small oscillation in the noise of the NMR spectrum can cause a large variation of PCA, which results in the form of irrelevant clustering. Robust Principal Component Analysis, a modified version of popular PCA method attempts to recover the low-rank matrix from the corrupted measurements. Though there are numerous versions of RPCA, Hubert et al. [18] proposed a method that can be described briefly as follows.

- At first, use singular value decomposition is used to deduce the data space.
- Next, for each data point a measurement of outlyingness is calculated. Out of all the n data points (suppose there are m smallest measurements), the covariance matrix, Σ_m is calculated. In addition, k number of principal components are chosen to retain.

- Finally, a subspace spanned by largest k number of eigenvalues corresponding to the k number of eigenvalues of the covariance matrix, Σ_m. Then all the data are projected onto the above sub space and the robust principal components are computed based on the location of the projected data points.

3.1.4 Random Forest

Random Forest is considered as one of the most popular and widely applied machine learning algorithms in data classification. Though the main use of Random Forest is for classification, it can be used as a useful regression model. Random Forest technique is an ensemble learning technique, proposed by Breiman [4]. Ensemble methods are considered as learning algorithms that builds a series of classifiers to classify a novel data point. This technique has found an answer to the overfitting problem available in individual decision trees (Fig. 2).

The entire classification process of Random Forest is achieved in a series of steps as described below.

- Suppose the number of training dataset contains N cases. A sub set from the above N is taken at random with replacement. These will be used as the training set to grow the tree.
- Assume there are M number of input variables. Then a number m, which is lower than M is selected and m number of variables from the collection of M is selected randomly. After that the best split of the selected m variables is selected to split the node. Throughout this process, m is kept as a constant.
- Without pruning, each tree is grown to the largest extent. The prediction of a new data point is by aggregating predictions.

Fig. 2 Random forest

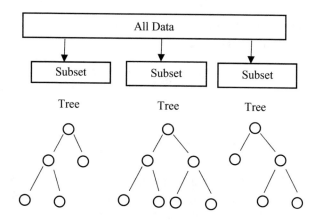

Table 1 True positive and false positive

Label	Meaning
TP-true positive	Positive items that are correctly classified as positives
TN-true negatives	Negative items that are correctly identified as negatives
FP-false positives	Negative items that are wrongly classified as positives
FN-false negatives	Positive items that are wrongly classified as negatives

3.2 Performance Indicators

As a means of quantifying the clarity of classification, following quantitative measurements are discussed. For this purpose, we consider several frequently used performance indicators, namely precision, specificity, sensitivity (recall), and F-Measure.

We adhere the following naming convention that is used to summarize the standard confusion matrix used in data classification (Table 1).

Let's assume P to be the total number of positive instances and N be the total number of negative instances. Then the performance indicators ca be defined as follows.

$$\text{Sensitivity} = \frac{FP}{P}$$

$$\text{Specificity} = \frac{FP}{N}$$

$$\text{Precision} = \frac{TP}{FP + TP}; \tag{4}$$

where P represents total number of positive (Class A), instances while N represents the total number of class B instances.

$$\text{F} - \text{Measure} = 2\left[\frac{Precision * Sensitivity}{Precision + Sensitivity}\right]$$

4 Dataset

In this study we utilize a dataset about online transactions using credit cards. This secondary dataset has been modified from the initial dataset, which contains credit cards' transactions by European credit cards holders within two days in September 2013. This dataset includes 29 features including time, amount, and the time duration of the transaction in seconds. Though it is interesting to know all the included attribute names in the dataset, due to the confidentiality issues the data do not disclose all the background information.

The response variable indicates whether there was an involvement of a fraud with the credit card transaction or not. Therefore, the feature (Class) takes value 1 to represent an occurrence of a fraud in the credit card transaction, while 0 represents the opposite. Due to the highly unbalanced nature of the dataset, the initial dataset was modified to prepare dataset of 1980 instances comprising of a nearly balanced dataset (997 fraud instances and 983 non-fraud instances).

5 Examining the Impact of Attribute Noise, and the Classification Technique on Classification Accuracy

5.1 The Noise

Data can be classified as attributes and class labels [49]. When noise is considered, it may either relate to independent variable (attribute) or to the dependent variable (class). Hence, if the noise exists in the data it can be classified as either attribute noise or class noise [23, 26, 42, 47, 49]. According to the literature findings, class noise brings more adverse effects than the attribute noise for classification. Despite the fact that class noise creates more damages than the attribute noise, it is believed that the latter is more complicate to handle. Therefore, we focus on studying attribute noise and the impact of it on data classification.

5.2 Attribute Noise, Classification Technique, and Classification Accuracy

Impact of attribute noise, and the classification technique on classification accuracy is tested in several phases. Four previously stated (Sect. 3.1) classification techniques were selected for this. As the first approach, credit card data was classified using the conventional SVM. In the second approach, PCA was used for data reduction before applying SVM. Third technique is exactly similar to the second, in which replacement of PCA by RPCA was the only difference. In the final technique, Random Forest is directly applied on the dataset similar to the first approach. After the application of each technique, sensitivity and the F-measure were calculated for the comparison.

Table 2 Average F-measure, classification methods, and noise

Method	5%	10%	15%	20%
SVM	0.48	0.92	0.96	1.82
PCA	1.56	0.33	0.62	1.20
RPCA	1.08	0.59	0.14	0.40
RForest	0.04	0.03	0.07	0.07

In the next phase random noise was introduced to the dataset in 5%, 10%, 15%, and 20% levels and F-measure was measured accordingly. Using the F-measures in each case, the percent change of F-measure was calculated. Table 2 summarizes this findings. In the implementation of algorithm, training and testing samples were generated according to the 70:30 ratio and performance indicators were calculated based on 100 randomly selected samples.

6 Skewness of the Data and Classification Accuracy

6.1 Skewness

Skewness provides a measure about the symmetry of the distribution. This measurement can be either positive or negative, in either case, skewness makes the symmetric distribution asymmetric. Transformation of skewed data into symmetric is often seen in data analysis. Though there is no strict set of guidelines regarding the type of transformation to use, following are some of the popular transformations. As Howell [17] suggested, if data are positively skewed $sqrt(X)$ is used. If the skewness is moderate, then the $log(X)$ can be used. Brown [5] stated the identification of skewness based on the Joanes and Gill [21] and Cramer [10] formulas as described below.

Consider a sample of n data points, $x_1, x_2, \ldots, x_n$. Then the method of moment coefficient of skewness is computed according the Eq. (5).

$$g = \frac{m_3}{m_2^{\frac{3}{2}}}; \text{ where } m_3 = \frac{\sum \left(x - \bar{x}\right)^3}{n}, m_2 = \frac{\sum \left(x - \bar{x}\right)^2}{n} \tag{5}$$

Using the above g, the sample skewness (G) is calculated according to the Eq. (6).

$$G = g\frac{\sqrt{n(n-1)}}{(n-2)} \tag{6}$$

Finally, the declaration of the skewness of the data is decided based on the value of Z_g. Here,

$$Z_g = \frac{G}{SES}; \text{ where SES, Standard Error of Skewness}$$

$$SES = \sqrt{\frac{6n(n-1)}{(n-2)(n+1)(n+3)}}$$

Therefore, we can classify the data as a negatively skewed if $Z_g < -2$. If $|Z_g| < 2$, then the distribution is either symmetric, negatively skewed or positively skewed. Finally if $|Z_g| > 2$, we can classify that the distribution is positively skewed.

6.2 Impact of Skewness on Classification

With the idea of quantifying the impact of skewness on classification accuracy, we generate data samples with noise levels 5%, 10%, 15%, and 20%. Each attribute in each sample is tested for the skewness as explained in Sect. 6.1. If the skewness is present, it is removed using an appropriate transformation. SVM, PCA, RPCA and RForest methods are applied afterwards. At the end, percent change of sensitivity and percent change of F-measure are calculated and these results can be seen in Table 3. In addition, Figs. 3 and 4 displays the observed quantitative measures graphically.

Table 3 Noise level, classification method, and the change of measures

Noise level	Method	% Change of sensitivity	% Change F-measure
5%	SVM	4.24	1.71
	PCA	1.20	0.44
	RPCA	0.20	0.32
	R FOREST	0.06	0.06
10%	SVM	1.68	1.32
	PCA	1.00	0.20
	RPCA	1.36	0.90
	R FOREST	0.08	0.08
15%	SVM	2.33	0.73
	PCA	0.14	0.55
	RPCA	2.40	0.87
	R FOREST	0.11	0.11
20%	SVM	2.23	0.50
	PCA	0.78	0.27
	RPCA	0.47	0.84
	R FOREST	0.05	0.05

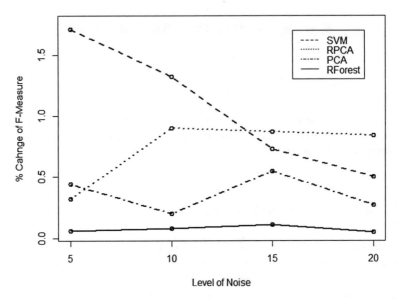

Fig. 3 Impact of skewness on F-measure

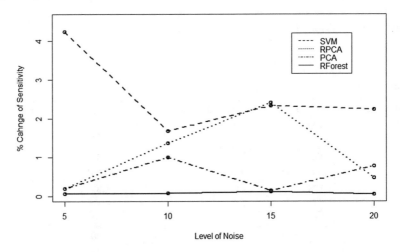

Fig. 4 Impact of skewness on sensitivity

7 Improving the Classification Accuracy When Noise Is Present

Improving the classification accuracy can be achieved in many ways. Appropriate model selection, purity of the data, and the use of efficient algorithms are some of the ways. Though much attention has not given, selection of appropriate sample sizes (training and testing samples) plays a major role in classification accuracy. With the

Table 4 Pseudo code of the algorithm for noise removal

```
Input : Data file with noise
Output: New data file after removing noise
   1.  For i=1 to (# of attributes)
   2.    For j=1 to (# observations)
   3.      For k=1 to (# bins)
   4.        bin the iᵗʰ attribute with selected bin
                sizes of 5, 10,15,25, and 50)
   5.        Calculate the mean and standard
                deviation of kᵗʰ bin
   6.        Convert each data points on bin k
                and attribute i into their
                corresponding Z-scores
   7.      End k
   8.    End j
   9.  End i
  10.  Summarizing the z-scores of each instance to
       a number and test for outliers using the
       method in section 5.1
  11.  Eliminate Outliers
```

presence of noise of the data, noise removal algorithm may be critically important for the classification accuracy. In this section, we aim to improve the classification accuracy by proposing a novel, but simple noise removal algorithm. This algorithm is tested using the credit card data and compare with one of the standard outlier detection method, based on the Cook's [9] distance.

7.1 A Simple Noise Removal Algorithm

As can be seen in the algorithm, the algorithm is executed in several steps. At first, the data in each attribute is binned according the selected bin sizes (5, 10, 15, 25, and 50). In the next step, data in each bin is converted into their corresponding sample z-scores by treating the entire bin as the sample. This process continues until the algorithm covers the entire dataset by taking each attribute at a time. After completing the z-score calculation for each attribute, standard outlier detection algorithm, as explained in Sect. 6.1 is applied. In the last step, outliers are removed from the dataset. The pseudo code of this proposed algorithm is displayed in Table 4.

This algorithm is implemented under each classification method on data with 5%, 10%, 15%, and 20% noise levels. Performance indicators are recorded before and after the implementation of the algorithm. Obtained results are shown in Figs. 5 and 6.

The effectiveness of this algorithm is compared with the Cook's distance approach. Logistic regression is fitted on the above data and Cook's distance is calculated on each data point. Then the data points are declared as unusual if

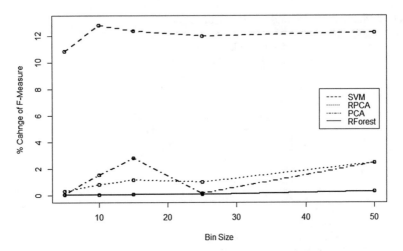

Fig. 5 Change (%) of F-measure, Bin Size, and Classification Method for Noise Level of 5%

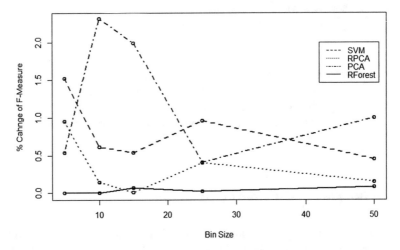

Fig. 6 Change (%) of F-measure, bin size, and classificationmethod for noise level of 15%

the Cook's distance is more than the ratio of 4/ *(total number of instances in the dataset)*. This is implemented for all the datasets of 5%, 10%, 15%, and 20% noise levels.

7.2   Impact of the Sample Size Ratio of Training and Testing on Classification Accuracy

It is evident that the classification accuracy would be biased if a large portion of data is selected to train the model by leaving very small portion to test the model. This may inflate standard deviations of the performance indicators. Therefore, identification of appropriate balance between the two sample sizes is very crucial. Therefore, it would be beneficial to understand how these ratio of samples impact the classification accuracy, when the data are mixed with noise.

We study this using nine sample ratios of training-test datasets. 90:10, 80:20, 70:30, 60:40, 50:50, 4:60, 30:70, 20:80, and 10:90 are the selected sample ratios. For each case, we consider datasets with 5%, 10%, 15%, and 20% noise levels. After simulating 100 instances, the average and the standard deviations of each measurement indicator were calculated at each noise levels. In addition, standard deviation and the mean values of F-measure are calculated.

7.2.1 Coefficient of Variation (CV)

When comparing two statistics, with different distributions for their standard deviations and means, the ratio between standard deviation (σ) and the mean (μ) values is considered as a better measurement. This measurement is considered as the Coefficient of Variation (CV), which can be calculated according to the Eq. 8. CV quantifies the dispersion of the statistics compared to its mean value.

$$CV = 100^* \frac{\sigma}{\mu} \tag{8}$$

After calculating above measurements, Figs. 7 and 8 display the fings.

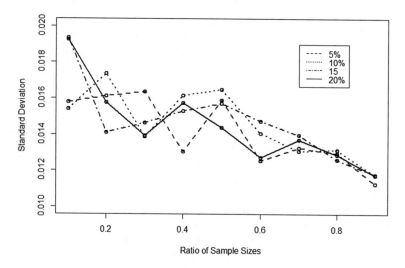

Fig. 7 Training: test ratio, noise levels and standard deviation of the F-measure

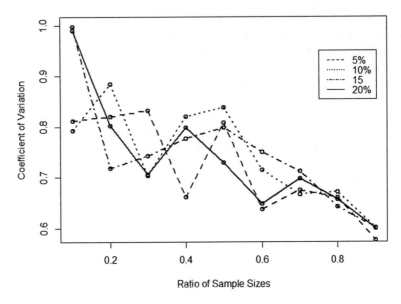

Fig. 8 Training: test ratio, noise levels and standard deviation of the F-measure

8   Results

Table 2 shows the average F-measure for each classification method at each noise
levels. Smaller values are indicating robustness of the model to the noise. According
to the Table 2 and outcomes of the Analysis of Variance (ANOVA), it is obvious
that there is a significant difference in the values of RForest method, compared to
the other counterparts $[F(3, 9) = 4.064, p = 0.0442]$. This means, Random Forest
method outperforms the rest of the alternatives across all the levels of noises.
Though SVM works better when there is less noise, with the increment of noise
RPCA works better than SVM.

Table 3 summarizes how skewness of the data influences the classification
accuracy, in the presence of noise. Even in this case, Random Forest method
shows the smallest change in both sensitivity and the F-measure. This indicates
that Random Forest method handles skewness of the data well. SVM shows that
it is not a robust classification technique for skewed data in comparison to the
other counterparts. Most importantly, the performance of SVM gets better with the
increment of noise levels. Out of PCA and RPCA, the latter performs better than
the former. This suggests that PCA is a better classification technique for noisy and
skewed data. This behavior is clearly explained by the Figs. 3 and 4.

Figures 5 and 6 display the relationship among the bin size, classification
technique, noise level and the classification accuracy. At 5% noise level, the highest
change of F-measure across all bin levels is recorded by SVM. Clearly, this is significantly different than other methods. This indicates that classification accuracy can

be significantly enhanced using the proposed algorithm, with SVM in particularly. As the Fig. 6 and other results indicate, PCA also performs better with this algorithm for higher levels of noise. All the methods, except the Random Forest show significant improvement of F-measures. This indicates that classification accuracy can be improved by implementing this noise removal algorithm. Furthermore, there exists a clear connection between the bin size and the classification accuracy. When the bin size is either 10 or 20, the highest performance can be achieved across all the classification techniques and all noise levels. ANOVA was conducted to evaluate the performance of capturing the noise using our proposed method. According to the findings there is a significant effect of bin size on the percentage of noise capturing at the $p < 0.05$ level for the five bin sizes [$F(4, 12) = 232.18, p = 0$]. When comparing effectiveness of the proposed algorithm and the Cook's distance method, there is significant evidence to claim that the proposed method captures higher percentage of noise than the alternative at $p < 0.05$, [$F(1, 6) = 35.01, p = 0.001$]. Finally, when studying the relationship between ratio of sizes for training-test datasets and the classification accuracy, Figs. 7 and 8 provide interesting findings. When the ratio is small, there is a higher variability of the performance indicator (F-measure) compared to its average. When the sample ratio is increased, the coefficient of variation decreases irrespective of the noise level. All the above stated outcomes were based on 100 iterations in each of the appropriate situation.

9 Conclusion and Future Work

In this empirical study we investigated the impact of attribute noise in the data on the classification accuracy. At first, we studied the influence of attribute noise on the classification accuracy. Secondly, we tried to enhance the power of data classification in the presence of noise. Hence, we collected a dataset about online transactions using credit-card, which is related to cyber-security. With this dataset, we classify whether a transaction has been involved a fraud or not. According to our findings, it is clear that classification accuracy is hindered by the presence of noise in the data. Furthermore, Random Forest method outperforms the other three classification techniques even in the presence of noise. SVM seems better than other two in the absence of Random Forest, but when the level of noise increases, RPCA performs better than the SVM. When testing the influence of skewness on data classification, we found that there is a direct impact from skewness on the classification accuracy. This influence affects differently on each technique. Among the selected techniques, Random Forest is robust even when data are skewed. Though SVM looks vulnerable to classify skewed data, when the noise level of the data is higher, SVM performs better. Further analysis shows that PCA can classify data well when the data are noisy and skewed. As a means of improving the classicization accuracy for noisy data, we proposed a simple noise removal algorithm. According to the obtained results, our algorithm significantly improve the classification accuracy. This was compared with the Cook's distance approach. According to the obtained results, our proposed algorithm shows better performances compared to the Cooks' distance approach. Further analysis indicates that there is a strong relationship between the selected bin size and the classification accuracy. Though this influence does not impact all the classification techniques uniformly, bin sizes 10 and 20 record higher performances than other bin sizes. At last, we studied about the appropriate ratio of sample sizes for both training and test datasets. According to the outcomes of this study, there is a strong connection between the ratio of datasets and the classification accuracy. When selecting appropriate sample sizes, one needs to pay attention about this ratios. As the results indicate, if the ratio is too small the variability of the performance indicator inflates. In cyber-security related data, enhancing the performance even in small amount will be advantageous. Though we have improved the classification accuracy significantly, this obtained outcomes motivate further research work to explore inclusion of novel classification techniques. In addition, future work will be conducted to see the influence other factors on the accuracy of data classification. This current study was conducted for the balanced data, therefore it would be interesting to extend this study for unbalanced data as well. Furthermore, an extension of the noise removal algorithm to address class noise will be beneficial too.

References

1. Abu-Nimeh, S., Nappa, D., Wang, X., & Nair, S. A comparison of machine learning techniques for phishing detection. In Proceedings of the anti-phishing working groups 2nd annual eCrime researchers summit, pp. 60-69, ACM. (2007).
2. Akbani R., Kwek S., and Japkowicz N.: "Applying support vector machines to imbalanced datasets," in Proceedings of the 15th European Conference on Machine Learning, pp. 39–50, (2004).
3. Beleites C., Neugebauer U., Bocklitz T., Krafft C., Popp J.: Sample size planning for classification models. Anal Chim Acta. Vol. (760), pp. 25–33, (2013).
4. Breiman L.: Random forests. Machine Learning, Vol. 45(1), pp. 5–32, (2001).
5. Brown S., Measures of Shape: Skewness and Kurtosis, https://brownmath.com/stat/shape.htm, (2008-2016)
6. Cao Y., Pan X., and Chen Y.: "SafePay: Protecting against Credit Card Forgery with Existing Card Readers", in Proc. IEEE Conference on Communications and Network Security, pp. 164–172, (2015).
7. Carrizosa, E., Martin-Barragan, B., Morales, D. R.: Binarized support vector machines. INFORMS Journal on Computing, Vol. 22(1), pp. 154–167, (2010).
8. Colas F., and Brazdil P.,"Comparison of SVM and Some OlderClassification algorithms in Text Classification Tasks", "IFIP International Federation for Information Processing", Springer Boston Volume 217, Artificial Intelligence in Theory and Practice, pp. 169–178, (2006).
9. Cook, R. D.: "Influential Observations in Linear Regression". Journal of the American Statistical Association. Vol. 74 (365), pp. 169–174, (1979).
10. Cramer, Duncan Basic statistics for social research: step-by-step calculations and computer techniques using Minitab. Routledge, London.; New York, (1997).
11. Cureton, Edward E, and Ralph B. D'Agostino. Factor Analysis, an Applied Approach. Hillsdale, N.J: L. Erlbaum Associates, (1983).

12. Díaz-Uriarte R., De Andres, S. A.: Gene selection and classification of microarray data using random forest. BMC bioinformatics, 7(1), p. 3, (2006).
13. Fabrice, R, Villa, N.: Support vector machine for functional data classification. Neurocomputing/EEG Neurocomputing, Elsevier, 69 (7–9), pp.730–742, (2006).
14. Guyon I.: A scaling law for the validation-set training-set size ratio, AT & T Bell Laboratories, Berkeley, Calif, USA, (1997).
15. Halouska S., Powers R.: Negative impact of noise on the principal component analysis of NMR data, Journal of Magnetic Resonance Vol. (178) (1), pp. 88–95, (2006).
16. Hickey R. J., "Noise modelling and evaluating learning from examples," Artif. Intell., vol. 82, nos. 1–2, pp. 157–179, (1996).
17. Howell, D. C. Statistical methods for psychology (6th ed.). Belmont, CA: Thomson Wadsworth, (2007).
18. Hubert, M., Rousseeuw, P. J., Branden, K. V.: ROBPCA: a new approach to robust principal components analysis, *Technometrics*, vol. 47, pp. 64–79, (2005).
19. Hwang, J. J., Yeh, T. C., Li, J. B.: Securing on-line credit card payments without disclosing privacy information. Computer Standards & Interfaces, Vol. 25(2), pp. 119-129, (2003).
20. Jayavelu D., Bar N.: A Noise Removal Algorithm for Time Series Microarray Data. In: Correia L, Reis L, Cascalho J, editors. Progress in Artificial Intelligence, vol. 8154. Berlin: Springer, pp. 152–62, (2013).
21. Joanes, D. N., Gill C. A.: "Comparing Measures of Sample Skewness and Kurtosis". The Statistician Vol. 47(1), pp. 183–189, (1998).
22. Kathiresan K., Vasanthi N. A., Outlier Detection on Financial Card or Online Transaction data using Manhattan Distance based Algorithm, International Journal of Contemporary Research in Computer Science and Technology (IJCRCST) Vol. 2(12), (2016).
23. Khoshgoftaar T., Hulse J. V.: Identifying noise in an attribute of interest. In ICMLA '05: Proceedings of the Fourth International Conference on Machine Learning and Applications (ICMLA'05), pp. 55–62, Washington, DC, USA, 2005. IEEE Computer Society. ISBN 0-7695-2495-8. doi: 10.1109/ICMLA.2005.39, (2005).
24. Lee C. C., Yoon J. W.: "A data mining approach using transaction patterns for card fraud detection", Seoul, Republic of Korea, pp. 1-12, (2013).
25. Liaw A., Wiener M.: Classification and Regression by Random Forest, R News, Vol. 2(3), (2002).
26. Liebchen G.: Data Cleaning Techniques for Software Engineering Data Sets. Doctoral thesis, Brunel University, (2011).
27. Maratea A., Petrosino, A.: Asymmetric kernel scaling for imbalanced data classification, in: Proceedings of the 9th International Conference on Fuzzy Logic and Applications, Trani, Italy, pp. 196–203, (2011).
28. Mennatallah A., Goldstein M., Abdennadher, S.: Enhancing oneclass support vector machines for unsupervised anomaly detection. In Proceedings of the ACM SIGKDD Workshop on Outlier Detection and Description pp. 8–15, (2013).
29. Miranda A. L., Garcia L. P., Carvalho A. C., Lorena A. C., "Use of classification algorithms in noise detection and elimination", Proc. 4th Int. Conf. Hybrid Artif. Intell. Syst., pp. 417–424, (2009).
30. Oja, E.: Principal components, minor components, and linear neural networks. Neural Networks, pp. 927–935, (1992).
31. Osborne, J. Notes on the use of data transformations. Practical Assessment, Research & Evaluation, 8(6). http://PAREonline.net/getvn.asp?v=8&n=6, (2002).
32. Redman, T.: Data Quality for the Information Age. Artech House, (1996).
33. Riem, A.: Cybercrimes of the 21st century: crimes against the individual—part 1, Computer Fraud and Security. Vol 6, pp. 13–17, (2001).
34. Rosenberg A.: "Classifying Skewed Data: Importance Weighting to Optimize Average Recall," Proc. Conf. Int'l Speech Comm. Assoc. (InterSpeech '12), (2012).
35. Sáez, J.A., Galar M., Luengo, J. et al. Analyzing the presence of noise in multi-class problems: alleviating its influence with the One-vs-One decomposition. Knowl Inf Syst 38: 179. doi: 10.1007/s10115-012-0570-1, (2014).

36. Sahin Y., Duman E.: "Detecting Credit Card Fraud by Decision Trees and Support Vector Machines", International Multi-conference of Engineers and computer scientists, (2011).
37. Scholkopf, B., Smola A. J.: Support Vector Machines and Kernel Algorithms, The Handbook of Brain Theory and Neural Networks. MIT Press, Cambridge, UK, (2002).
38. Seo S.: Masters thesis. University of Pittsburgh; Pennsylvania: A review and comparison of methods for detecting outliers in univariate data sets, (2006).
39. Shahi A., Atan R. B., Sulaiman M. N.: Detecting effectiveness of outliers and noisy data on fuzzy system using FCM. Eur J Sci Res 36: pp. 627–638, (2009).
40. Siddiqui F., and Ali, Q. M.: Performance of non-parametric classifiers on highly skewed data, Global Journal of Pure and Applied Mathematics. ISSN 0973-1768 Vol. 12(2), pp. 1547–1565, (2016).
41. Tang L., Liu H.: Bias analysis in text classification for highly skewed data. In ICDM '05: Proceedings of the Fifth IEEE International Conference on Data Mining, IEEE Computer Society, pp. 781–784, (2005).
42. Teng M. C.: Combining noise correction with feature selection. pp. 340–349, (2003).
43. Vapnik V., *The Nature of Statistical Learning Theory*. Springer-Verlag, ISBN 0-387-98780-0, (1995).
44. Wang, Bin, et al. "Distance-based outlier detection on uncertain data." Ninth IEEE International Conference on Computer and Information Technology, 2009. CIT'09. Vol. 1. IEEE, (2009).
45. Wei X., and Yuan L.: "An Optimized SVM Model for Detection of Fraudulent Online Credit Card Transactions," International Conference on Management of e-Commerce and e-Government, 2012.
46. Xiong H., Pandey G., Steinbach M, Kumar V.: "Enhancing data analysis with noise removal," IEEE Trans. Knowl. Data Eng., Vol. 18(3), pp. 304–319, (2006).
47. Yoon K., Bae D.: A pattern-based outlier detection method identifying abnormal attributes in software project data. Inf. Softw. Technol., Vol. 52(2), pp. 137–151. ISSN 0950-5849. (2010).
48. Zhou X., Zhang Y., Hao S., Li S., "A new approach for noise data detection based on cluster and information entropy." The 5th Annual IEEE International Conference on Cyber Technology in Automation, Control, and Intelligent Systems, (2015).
49. Zhu X., Wu X., Class noise vs. attribute noise: a quantitative study, Artificial Intelligence Review Vol. 22 (3). pp.177–210, (2004).
50. Zhu, X., Wu X., Yang, Y.: Error Detection and Impact-sensitive Instance Ranking in Noisy Datasets. In Proceedings of 19th National conference on Artificial Intelligence (AAAI-2004), San Jose, CA. (2004).

Part II
Invited Chapters

Learning from Loads: An Intelligent System for Decision Support in Identifying Nodal Load Disturbances of Cyber-Attacks in Smart Power Systems Using Gaussian Processes and Fuzzy Inference

Miltiadis Alamaniotis and Lefteri H. Tsoukalas

Abstract The future of electric power is associated with the use of information technologies. The smart grid of the future will utilize communications and big data to regulate power flow, shape demand with a plethora of pieces of information and ensure reliability at all times. However, the extensive use of information technologies in the power system may also form a Trojan horse for cyberattacks. Smart power systems where information is utilized to predict load demand at the nodal level are of interest in this work. Control of power grid nodes may consist of an important tool in cyberattackers' hands to bring chaos in the electric power system. An intelligent system is proposed for analyzing loads at the nodal level in order to detect whether a cyberattack has occurred in the node. The proposed system integrates computational intelligence with kernel modeled Gaussian processes and fuzzy logic. The overall goal of the intelligent system is to provide a degree of possibility as to whether the load demand is legitimate or it has been manipulated in a way that is a threat to the safety of the node and that of the grid in general. The proposed system is tested with real-world data.

1 Introduction

The application of digital control system technologies and sensor networks for monitoring purposes in critical power facilities is a topic of high interest [1]. The use of digital technologies offers significant advantages that include reduction in the purchase and maintenance costs of facility components and equipment, and a significant reduction in the volume of hardware deployed throughout the facility [2].

M. Alamaniotis (✉) • L.H. Tsoukalas
Applied Intelligent Systems Laboratory, School of Nuclear Engineering,
Purdue University, West Lafayette, IN, USA
e-mail: malamani@ecn.purdue.edu; tsoukala@ecn.purdue.edu

© Springer International Publishing AG 2017
I. Palomares et al. (eds.), *Data Analytics and Decision Support for Cybersecurity*,
Data Analytics, DOI 10.1007/978-3-319-59439-2_8

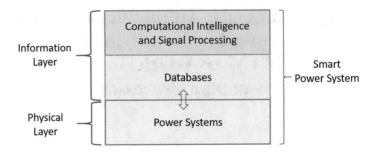

Fig. 1 The notion of smart power systems that are comprised of two layers

The basic mission of power systems is the nonstop delivery of generated electricity to connected consumers. Delivery systems have been equipped with several monitoring, prognostics and diagnostics tools, as well as several redundant systems and mechanisms [3]. It should be noted that redundancy refers to availability of more than required systems and mechanisms that perform the same operation. Redundancy allows retaining of the normal operation of the power system in case a system or mechanism fails, given that the rest will compensate for it. Thus, power systems exhibit high resiliency and fault tolerance behavior in fulfilling their mission of 24/7 power delivery. The advent of computers and internet connectivity added one more factor that should be taken intro serious consideration: cyber security [4]. The advantages offered by the ever increasing use of computer and communication technologies (ICT) in power system operation, come at a cost of new vulnerabilities in system security. What was formerly a physically isolated system is now open to unauthorized access and manipulation control through potential vulnerabilities posed by ICT use [5]. For instance, the supervisory control and data acquisition (SCADA) systems consist of the backbone of the overall power system monitoring and subsequent decision-making processes. Hence, despite their nowadays practical benefits, SCADA systems present serious targets for cyber-attackers.

The advent of smart power technologies is expected to couple data technologies with conventional power systems in ways that optimizes energy consumption and minimizes loses [6]. The notion of smart power systems is visualized in Fig. 1, where it can be observed that the smart power system is a form of a cyber-physical system [7, 8]. The information (i.e., cyber) layer is comprised of two modules. The first module includes the intelligent systems and signal processing tools and the second module the databases. Those two modules are complementary; the intelligent tools utilized stored information to make management decisions pertained to power system operation [9]. The second layer is the physical layer that contains the physical components of the power system.

Given that smart power systems are heavily dependent on data and information processing tools as it is depicted in Fig. 1, cybersecurity is a major issue that cannot be overlooked. Every asset of smart power systems may be compromised and subsequently be transformed into a vehicle for conducting a cyber-attack [10].

In contrast to physical attacks, the nature of cyber-attacks imposes significant difficulties in detecting, isolating and mitigating the consequences of cyber-attacks. Additionally, estimation of a cyber-attack impact is a difficult task given that full detection of all parts that have been affected by the attack, if possible, may take a significant amount of time [11]. Full recovery is also a time-consuming process. The cyber-attack in Ukrainian power grid on December 2015 presents an example in which computer control is not fully restored yet [12, 13]. In addition, cyberattacks may be used as a proxy for a subsequent physical attack in critical energy facilities. For instance, manipulation of digital access may allow a person to get access to areas of nuclear material disposal that may be used for nefarious purposes [14].

Detection of cyber-attacks and their consequences in operation of smart power systems may be characterized as a multi-stage task. This is true since these systems contain the physical layer (power infrastructure) as well as the cyber layer (information technologies) [15–17]. Furthermore, the correlations among the two layers as well as the associations between components of the same layer makes detection of the source of an attack a very challenging task. For example, a cyber-attack's objective may be to cause a disturbance in a substation and thus it attempts to take control over the smart relays that reside in that substation [18, 19]. Hacking the operator's computer may perform this type of attack, that is, stealing its login credentials, and then through the operators' computer and substation communication, the hacker assumes control of the protective relays [20].

The technological success and prevalence of smart power technologies may facilitate the development of intelligent scientific methods and tools for achieving effective grid cybersecurity from various perspectives. To that end, research efforts will contribute in detecting cyberattacks utilizing power consumption and forecasting information [21]. Intelligent monitoring tools may be connected to a node of the power grid infrastructure and independently of the wide area monitoring system (WAMS) obtain and process node data [22]. Hence, nodal analysis of power information may reveal discrepancies between expected and actual consumption at that node.

In this work, we study the suitability of load demand information, and we present "learning from loads" as a method for detecting disturbances in power systems nodes. In particular, we introduce an intelligent system as a means for detecting load disturbances at the nodal level. We theorize that an intelligent system will leverage the non-linear dynamics of the load demand of a very short term horizon, and that will be adequate to distinguish malicious disturbances from normal ones. Load demand at the nodal level results from the aggregation of individual consumer demands. Individual demand exhibits high variance while the aggregation of multiple demands provides a signal that is smoother. In simple terms we can say that the greater the number of individual consumers or nodes, the higher the smoothing of the demand signal. Thus, aggregation poses difficulties in identifying load disturbances because aggregation may mask the disturbances on a single individual node. In our work, we implement an intelligent system that utilizes the synergism of Gaussian process regression (GPR) and fuzzy logic. The GPR is used for prediction while the fuzzy logic tool makes inferences regarding decisions as to whether the demand is legitimate or it has been manipulated by an attacker.

The remainder of this chapter contains five sections. Section 2 briefly introduces GPR and fuzzy logic inference, while Sect. 3 introduces smart power system and in particular Energy internet. Section 4 describes the proposed intelligent system. Section 5 presents the results obtained by applying the intelligent system to a set of load demand data. At the end, Sect. 6 concludes and summarizes the salient points of our approach.

2   Background

This section is dedicated in briefly presenting the theoretical foundation underlying our approach. In particular, the Gaussian Process Regression in the context of kernel machines will be introduced followed by a short discussion on fuzzy logic inference. The theory presented in this section will foster the ground for understanding the proposed intelligent system.

2.1   Gaussian Process Regression

Machine learning has been identified as one of the pillars in developing efficient data analytics methods and decision support systems. One of the preeminent areas of machine learning is the class of non-parametric methods called *kernel machines*.

A kernel machine is any analytical model that is a function of a kernel [23]. A kernel (a.k.a., *kernel function*) is any valid analytical function that takes the following form:

$$k(x_1, x_2) = \varphi(x_1)^T \cdot \varphi(x_2) \tag{1}$$

where, $\varphi(x)$ is called the basis function, and x_1, x_2 are input values. The inputs to a kernel may be either both scalar or vector values of equal length. Their range of values depends on the problem under study, while the kernel output represents the similarity between the input values. In general, selection of the basis function falls within responsibilities of the modeler and depends on the specifics of the application at hand [24]. For instance, the simplest basis function is $\varphi(x) = x$ and therefore the respective kernel takes the form given below:

$$k(x_1, x_2) = x_1^T \cdot x_2 \tag{2}$$

which is simply known as the linear kernel. It should be noted that formulation of models using kernels whose form can be determined by the modeler is called the *kernel trick* and finds wide use in data analytics and pattern recognition applications [23].

The class of kernel machines contains the Gaussian processes (GP). Gaussian processes may be modeled as a function of a kernel. In particular, the kernel enters into the Gaussian process formulation through its covariance function. A kernel-modeled Gaussian process can be used either in classification or regression problems. In the latter case, it is identified as Gaussian process regression [23].

Likewise Gaussian distribution, a Gaussian process is identified via its two parameters, namely, the mean function and the covariance function denoted by $m(x)$ and $C(x^T,x)$ respectively. Thus, we get [23]:

$$GP \sim N\left(m(x), C\left(x^T, x\right)\right). \tag{3}$$

Derivation of the GPR framework has as a starting point Eq. (3) where we set

$$m(x) = 0, \tag{4}$$

and

$$C\left(x^T, x\right) = k\left(x^T, x\right). \tag{5}$$

Selection of $m(x)=0$ is a convenient choice, while the covariance function is replaced by a kernel function [23]. In that way, GPR is transformed into a kernel machine that may be used in regression problems.

The above GPR formulation necessitates the availability of datasets, i.e., known targets t for known inputs x (in other words training datasets). The size of the training population is denoted as N. Thus, we assume that we have N known data points, which are consolidated in a matrix x_N, and we denote a new unknown one as x_{N+1}. The respective target associated with x_{N+1} is denoted as t_{N+1}. It should be noted that GPR considers the joint distribution between the N training data points and the unknown x_{N+1} to be s a Gaussian distribution. Based on that assumption, it has been shown in [25, 26] that GPR provides a prediction interval over the target t_{N+1} that is denoted in the form of a predictive distribution. That predictive distribution is Gaussian with a mean and covariance function given by:

$$m\left(x_{N+1}\right) = \mathbf{k}^T \mathbf{C}_N^{-1} \mathbf{t}_N \tag{6}$$

$$\sigma^2\left(x_{N+1}\right) = k - \mathbf{k}^T \mathbf{C}_N^{-1} \mathbf{k} \tag{7}$$

with $\mathbf{C}_N$ being the covariance matrix among the N training data, $\mathbf{k}$ being a vector of covariances between the input x_{N+1} and the N training data, and k the scalar value taken as $k(x_{N+1}, x_{N+1})$ [23].

Thus, kernel selection should be done carefully and with respect to the respective GPR output. Overall, kernel-based GPR offers flexibility in prediction-making; the modeler is able to select a kernel among existing ones or compose a new kernel. Hence, the kernel form can be tailored to the specifics of the prediction problem, allowing the modeler to have flexibility in the way he builds his prediction model.

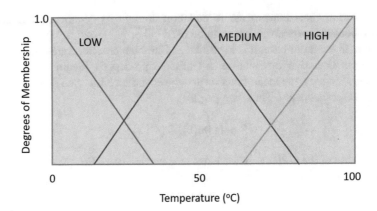

Fig. 2 Example of fuzzy sets pertained to temperatures [0, 100] °C

2.2 Fuzzy Logic Inference

Fuzzy logic is a branch of artificial intelligence that finds application in data analytics under vagueness. The strength of fuzzy logic is the representation of vaguely expressed knowledge via the use of linguistic descriptions. Each linguistic description is modeled by a fuzzy set [27], which may be seen as an extension of the classical set theory.

In classical set theory, an object clearly belongs to a set or not. In other words, crisp sets follow a digital logic, where an object either belongs to a set and take a value 1 for that set, or does not and takes a value of 0. A fuzzy set defined over a universe of discourse U is a set whose objects in U belong to the set up to a degree; the degree values are taken in the interval [0 1]. For instance, an object may belong to a set A with a degree of membership equal to 0.3 while at the same time it belongs to the set B with a degree of membership equal to 0.8. It should be noted that crisp sets may be seed as fuzzy sets whose objects take degrees of membership either 1 or 0.

Evaluation of the degrees of membership is performed via the use of the membership function. A membership function assigns to each object of the universe of discourse a number in the interval [0 1] that denotes the degree to which this object belongs to the set. Figure 2 shows an example where the temperatures [0, 100] °C are spanned by three fuzzy sets.

The use of fuzzy sets facilitates implementation of fuzzy inference mechanisms called fuzzy inference. The fuzzy inference mechanism is expressed via the use of *IF..THEN* rules, where the *IF* part contains the conditions, and the *THEN* part contains the consequences. For instance, two simple rules may have the following form:

IF x is A, THEN y is B

IF x is C, THEN y is D

where *A*, *B*, *C* and *D* are fuzzy sets. By using the sets in Fig. 2, an example of a fuzzy controller for a heating system might have the following rules:

IF temperature is LOW, THEN heater intensity is HIGH

IF temperature is MEDIUM, THEN heater intensity is MEDIUM

IF temperature is HIGH, THEN heater intensity is LOW

where, *temperature*, and *heater intensity* are called fuzzy variables.

Overall, a fuzzy inference system is comprised of several rules of the above form. In fuzzy inference systems one or more rules may be fired and therefore a fuzzy output is taken [28]. In that case, a defuzzifying step is utilized to obtain a single output value [27, 28]. One of the widest used methods of defuzzification is the *center of area* (COA) method whose analytical formula is given below [27]:

$$y^* = \frac{\sum_{n=1}^{N} x_n \mu_A (x_n)}{\sum_{n=1}^{N} \mu_A (x_n)} \tag{8}$$

where y^* is the defuzzified output value, and $\mu_A(x_n)$ is the degrees of membership of the input x_n to fuzzy set *A*. The COA may be seen as the weighted average value of the obtained fuzzy output; in practice it computes the area below the fuzzy output and finds its mean value.

Fuzzy inference has found various applications and fuzzy systems for developing decision support systems are under development. The interested reader is advised to check Ref. [27] for details on fuzzy inference, where several illustrative examples are described in detail.

3 Smart Power Systems: The Energy Internet Case

This section includes a short introduction of the concept of price-directed smart power systems and in particular the principle idea of an Energy Internet as proposed by the Consortium for the Intelligent Management of Electric Grid (CIMEG) [28]. The goal of this section is to provide the general framework of smart power systems and show the general environment in which the proposed intelligent system is of practical use.

The Energy Internet exploits advancements of information networks to leverage energy utilization. Crucial in implementing an Energy Internet are energy anticipation at consumer and at nodal level, smart meter negotiations, and determination of real-time price signaling [30]. The general block diagram of a smart power system developed by CIMEG is presented in Fig. 3. CIMEG models the power grid as a demand-driven system. It adopts a bottom up approach to determine the global state of grid health. To that end, CIMEG introduced the notion of a Local Area Grid

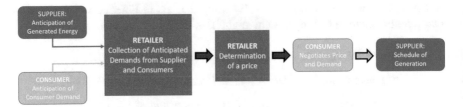

Fig. 3 General framework of Smart Power Systems implemented as an Energy Internet [30]

(LAG) that is characterized as the clustering of several different consumers. A LAG is responsible for keeping its own stability by taking the necessary actions when appropriate [6].

We observe in Fig. 3 that the individual consumers provide an anticipation of their demand for a specific ahead of time horizon. At the same time, suppliers also anticipate their generated electrical energy for the same time horizon. Both parties, i.e., consumers and suppliers forward their anticipated signals to the retailers. Subsequently, the retailer collects the anticipations and by utilizing price elasticity models determines a price for each of the consumer. Then, the consumer has the opportunity to negotiate the price with the retailer by altering its initial anticipated demand [31]. Through negotiations, retailer and consumer come to an agreement. Once all consumers make an agreement with the retailer, then the generator schedules the generated energy to meet final demand.

Iteration of the above process may take place at every system operational cycle. Details on smart power and Energy Internet may be found in references [6, 32, 33].

4 "Learning From Loads" Intelligent System

Having introduced the framework of smart power (energy internet) this section presents the "learning from loads" system to detecting nodal load disturbances for enhancing cybersecurity in this framework. In the following subsection, the proposed methodology as well as the attack scenarios pertained to load disturbances are presented.

4.1 Intelligent System Architecture

In this section, the intelligent system that makes decisions pertained to nodal load with respect to cybersecurity is presented. It should be noted that the proposed intelligent system exploits current as well as future information via the use of anticipation functions. Anticipation will allow the system to evaluate its future states compared to what it has seen so up to this point. To that end, a learning from load demand approach is adopted, and subsequently anticipates the future demand.

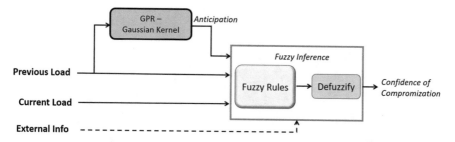

Fig. 4 Block diagram of the proposed learning from loads intelligent system

Anticipation is utilized to evaluate the actual demand and make inferences whether data have been manipulated or not.

The block diagram of the proposed system is depicted in Fig. 4. We observe that there are two paths. The first path analyzes recorded data and anticipates future load, while the second path utilizes the current load. In addition, to the above paths, we may observe in Fig. 3 that there is a subsidiary path that contains external information pertained to that specific node. For instance, the price of electricity at current time.

Load anticipation is performed using kernel modeled Gaussian process regression, which was introduced in the previous section. The GPR is equipped with a Gaussian kernel whose analytical form is given below:

$$k(x_1, x_2) = \exp\left(-\frac{\|x_1 - x_2\|^2}{2\sigma^2}\right) \tag{9}$$

where σ^2 is a kernel parameter whose value is being determined in the training phase of the algorithm. The GPR model is trained on previous recorded load demands. Once the training is finished the GPR is utilized for making prediction of the future demand. Therefore, we observe that this is the first point at which our system learns from loads (i.e., past loads).

Further, we observe in Fig. 3 that the information, anticipated, previous and current loads are fed to a fuzzy inference system [34]. In addition, the external information is also directly fed to the fuzzy inference system. Overall, it should be noted that the available information is forwarded to the fuzzy inference.

The fuzzy inference system is comprised of two parts as Fig. 3 depicts. The first part contains the fuzzy rules utilized for making inference. The fuzzy rules are predetermined by the system modeler and they take the form of IF.. THEN rules as shown in Sect. 2 as well. The left-hand side of the rules, i.e., conditions, may be refer to anticipated load or current load. In particular, the rules for anticipated load have the following general form:

IF Future Load will be $A_F(t)$, THEN Confidence of Compromise is $B(t)$

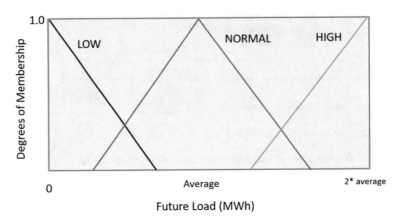

Fig. 5 Fuzzy sets for fuzzy variable *Future Load*

where the fuzzy variable is *Future Load* and A_F denotes the fuzzy values (see Fig. 5) while the rules for current load have the form:

IF Current Load is A(t), THEN Confidence of Compromise is B(t).

with Current Load being the fuzzy variable in that case (see Fig. 6).

In addition, we defined a new fuzzy variable which is equal to the absolute difference between the Anticipated Load and the Actual Load of the same time interval, i.e.:

$$Difference = | \, Anticipated \, Load - Actual \, Load \, |$$

where the variable Difference is actually the fuzzified value obtained by subtraction of anticipated and actual load. This fuzzy variable will allow to measure how much was the inconsistency between the anticipated and the actual load over the same time interval. It should be noted that the actual load is different from the current load. The current load exhibits the current load demand while the actual load variable refers to the actual demand for the time interval that the anticipation was made. This is a fundamental inference variable given that the offset between anticipation and actual may "reveal" the presence of a threat or not.

Lastly, we model the current prices of electricity as external information. The presented intelligent system aspires to be deployed in smart power environment (energy internet) and therefore price will play an important role in justifying load disturbances. The price will show the willingness of the consumer to change its consumption. For instance, an attacker will increase the load demand no matter how high the prices are; that can be used as an indicator to identify potential attackers. The rules pertained to price have the following form:

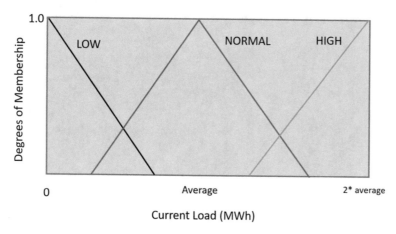

Fig. 6 Fuzzy sets for fuzzy variable *Current Load*

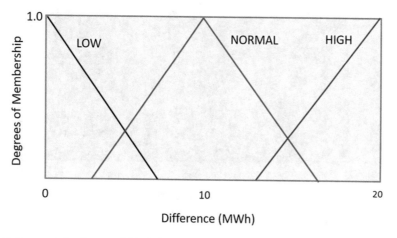

Fig. 7 Fuzzy sets for fuzzy variable *Difference*

IF Price is A(t), THEN Confidence of Compromise is C(t).

Additionally, it should be noted that as an external information we consider: (1) the average current load that we observe for that node, and (2) the average anticipated load that the GPR provides. These values may be found by keeping record of previous values.

Figures 5, 6, 7, and 8 provide the fuzzy sets of the fuzzy variables *Future Load, Current Load, Difference,* and *Price* respectively, while Fig. 9 depicts the fuzzy sets of the *Confidence of Compromise* variable (the confidence scale is in interval [0 1]). The fuzzy inference overall was comprised of 20 fuzzy rules and was implemented in Matlab software.

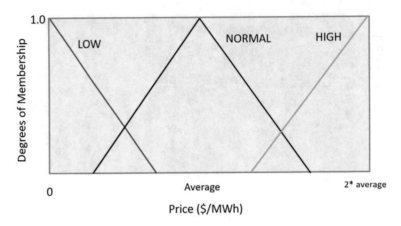

Fig. 8 Fuzzy sets for fuzzy variable *Price*

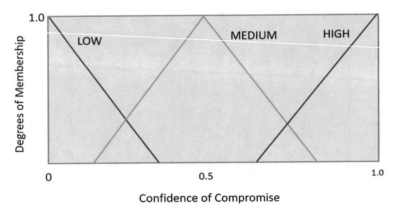

Fig. 9 Fuzzy sets for fuzzy variable *Current Load*

The second part of the fuzzy inference systems includes the defuzzification part. The defuzzification uses the center of area method [27] whose form is given in Eq. (8). The defuzzification provides a crisp output that is also the final output of the system. The output value is the confidence of compromise pertained to load disturbances for that node. The value of the confidence may support the power system operators and security engineers to make decisions whether there is cyber-attack pertained to load demand.

4.2 Type of Threats for Nodes of Smart Power Systems

The experience of cyberattack in the Ukrainian power system in December 2015, proved that the tactical preparation of the cyber-attackers maybe well defined and planned over a long time. Furthermore, the attackers may show patience, perform long term reconnaissance and follow a carefully planned series of actions. Given that smart power systems are heavily dependent on the utilization of information, the cyberattacks may also involve a significant deal of grid congestion by increasing nodal load demand.

The scenario that we examine in this work is the following:

- A cyberattacker wants to cause a blackout in a part of the grid. One way to achieve this is by destabilizing one of the grid nodes. Destabilzation can be done by increasing the demand beyond the capacity limits of the node. Therefore, the node will become nonoperational and the power delivery will fail at that point. In addition, if that node is at the backbone of the grid then it may propagate the problem to the rest of the grid.
- A simple way to cause this type of attack is to compromise the intelligent meters of several consumers that are connected to that node. As we noted, in the vision of Energy Internet the consumer negotiates with the retailer. Negotiations may be undertaken by the attacker with the consumer having no knowledge about it. Overall, by taking into control several of the consumers the attacker may cause a blackout in the node.

In the next section, the above scenario will be employed in a real case scenario. Nodal demand from real-world datasets will be used as our test scenario. The intelligent system will be utilized to analyze the load (current and anticipated) and output a confidence value whether there is manipulated increase in load demand.

5   Results

In this section, we test the proposed intelligent system on a set of real world data. The datasets contain the hourly values of loads and energy prices within the smart grid system, for one day before the targeted day. For visualization purposes the actual demand signal is given in Fig. 10.

The goal is detect whether the security status of a node in the grid was compromised. We assume that load demands are recorded every hour. Therefore, the GPR model is utilize for anticipating the load for the next hour. The predicted signal is depicted in Fig. 11. It should be noted that the training of the GPR was performed by utilizing the demand data of one and two days before the targeted day.

In our work, we tested our intelligent system on the following scenarios:

(1) No compromise has occurred.
(2) From 12.00 pm to 15.00 pm, the node is compromised and 10% increase in demand is presented
(3) From 12.00 pm to 15.00 pm, the node is compromised and 50% increase in demand is presented

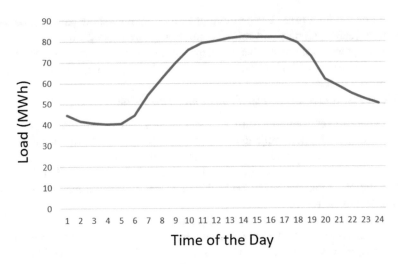

Fig. 10 Actual Demand signal for the tested day

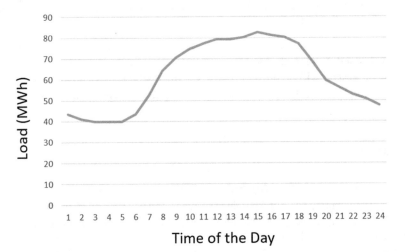

Fig. 11 Anticipated by GPR signal for the tested day

(4) From 12.00 pm to 15.00 pm, the node is compromised and 75% increase in demand is presented
(5) From 12.00 pm to 15.00 pm, the node is compromised and 90% increase in demand is presented.

The presented intelligent system is applied in the above scenarios and a degree of confidence per hour is taken. The respective degrees of confidence are given in Table 1. We observe in Table 1, that in the majority of the cases the intelligent system provides low confidence with regard to confidence of compromise. The value is not zero, mainly because of the uncertainty in load prediction and the volatility of

Table 1 Results obtained for the five tested scenarios

| Time | Confidence of compromise | | | | |
	Scenario 1	Scenario 2	Scenario 3	Scenario 4	Scenario 5
12.00 am	0.20	0.20	0.20	0.20	0.20
1.00 am	0.20	0.20	0.20	0.20	0.20
2.00 am	0.20	0.20	0.20	0.20	0.20
3.00 am	0.20	0.20	0.20	0.20	0.20
4.00 am	0.20	0.20	0.20	0.20	0.20
5.00 am	0.20	0.20	0.20	0.20	0.20
6.00 am	0.20	0.20	0.20	0.20	0.20
7.00 am	0.20	0.20	0.20	0.20	0.20
8.00 am	0.20	0.20	0.20	0.20	0.20
9.00 am	0.20	0.20	0.20	0.20	0.20
10.00 am	0.20	0.20	0.20	0.20	0.20
11.00 am	0.20	0.20	0.20	0.20	0.20
12.00 pm	0.20	0.31	0.74	0.80	0.88
1.00 pm	0.20	0.35	0.74	0.82	0.90
2.00 pm	0.20	0.31	0.72	0.80	0.88
3.00 pm	0.20	0.31	0.72	0.80	0.88
4.00 pm	0.20	0.20	0.20	0.20	0.20
5.00 pm	0.20	0.20	0.20	0.20	0.20
6.00 pm	0.20	0.20	0.20	0.20	0.20
7.00 pm	0.20	0.20	0.20	0.20	0.20
8.00 pm	0.20	0.20	0.20	0.20	0.20
9.00 pm	0.20	0.20	0.20	0.20	0.20
10.00 pm	0.20	0.20	0.20	0.20	0.20
11.00 pm	0.20	0.20	0.20	0.20	0.20

prices, as well as in the defuzzificaton method. The center of area method does not get the extreme values (0 and 1). However, a low confidence is an adequate sign to make the operator believe that there is no cyberattack. Though there is no specific decision threshold, any confidence value above 0.7 denotes a certain occurrence of node compromise.

Regarding the scenarios that we have intentionally increased the demand following the goal of the attack (to make the node blackout), we observe the result in the shaded area at the center of Table 1. We observe in scenario 2 that the confidence slightly increases. This increase is very small and therefore the operator may ignore it. However, the amount of load that was compromised is only 10% and thus we assume that this is not enough to cause serious problem to that node. Overall, this compromise may not be detected; however, the goal of the system is to support stakeholders in decision-making tasks, rather than making the actual decisions.

With regard to scenarios 3,4 and 5 the confidence increases above 0.7 and therefore we can consider that a cyberattack has been detected with high confidence. Such high confidence is a serious indicator that something is wrong, and the demand increased beyond the regular one. Therefore, it safe to conclude that the intelligent system presented detects the threat in those cases with high confidence and hence, it supports the correct decision by the operator.

6　Conclusions

Smart power systems are systems that integrate power with information, and as a result may become targets for cyberattackers. The importance of the power grid for innumerable everyday activities and modern life makes the defense of the grid mandatory. In addition, recent recorded attacks showed that cyberattacks may be carefully planned and not just opportunistic cases for attackers to show off. Therefore, every aspect of smart power systems should be secured.

Intelligent systems offer new opportunities for implementing new decision support and data analysis methods that mimic the way of human system operators. In this work, we examined the case in which the attacker plans to congest a power grid node by increasing the demand and causing a blackout. An intelligent system that learns from load signals by utilizing GPR and fuzzy inference was developed and applied on a set of five different scenarios. The results were encouraging: the higher the "compromised demand" the higher the degree of confidence that the system is being compromised provided by our system. Therefore, our system shows a high potential for its deployment into smart power systems, and in particular for an Energy Internet scenario is high.

Future work will contain two main directions. In the first direction, we will explore the use of other kernel function beyond the Gaussian kernel for prediction making in the GPR model. In particular, we will apply a variety of kernels on load data from coming from different nodes, and record their performance. Analysis of records will be used to develop a system for indicating the best kernel for each node. In the second direction, we will extensively test our intelligent systems in a higher variety of data including data of nodes from different geographical areas, and different assembly of customers.

References

1. Wood, A. J., & Wollenberg, B. F. (2012). *Power generation, operation, and control.* John Wiley & Sons.
2. Amin, S. M., & Wollenberg, B. F. (2005). Toward a smart grid: power delivery for the 21st century. *IEEE power and energy magazine, 3*(5), 34–41.
3. Han, Y., & Song, Y. H. (2003). Condition monitoring techniques for electrical equipment-a literature survey. *IEEE Transactions on Power delivery, 18*(1), pp. 4–13.

4. Li, S., Li, C., Chen, G., Bourbakis, N. G., & Lo, K. T. (2008). A general quantitative crypt-analysis of permutation-only multimedia ciphers against plaintext attacks. *Signal Processing: Image Communication, 23*(3), pp. 212–223.

5. Ramsey, B. W., Stubbs, T. D., Mullins, B. E., Temple, M. A., & Buckner, M. A. (2015). Wireless infrastructure protection using low-cost radio frequency fingerprinting receivers. *International Journal of Critical Infrastructure Protection, 8*, 27–39.

6. Alamaniotis, M., Gao, R., & Tsoukalas, L.H., "Towards an Energy Internet: A Game-Theoretic Approach to Price-Directed Energy Utilization," in *Proceedings of the 1st International ICST Conference on E-Energy*, Athens, Greece, October 2010, pp. 3–10.

7. Alamaniotis, M., Bargiotas, D., & Tsoukalas, L.H., "Towards Smart Energy Systems: Application of Kernel Machine Regression for Medium Term Electricity Load Forecasting," *SpringerPlus – Engineering*, Springer, vol. 5 (1), 2016, pp. 1–15.

8. Karnouskos, S. (2011, July). Cyber-physical systems in the smartgrid. In *Industrial Informatics (INDIN), 2011 9th IEEE International Conference on* (pp. 20-23). IEEE.

9. Alamaniotis, M., & Tsoukalas, L.H., "Implementing Smart Energy Systems: Integrating Load and Price Forecasting for Single Parameter based Demand Response," *IEEE PES Innovative Smart Grid Technologies, Europe (ISGT 2016)*, Ljubljana, Slovenia, October 9-12, 2016, pp. 1–6.

10. Beaver, J. M., Borges-Hink, R. C., & Buckner, M. A. (2013, December). An evaluation of machine learning methods to detect malicious SCADA communications. In *Machine Learning and Applications (ICMLA), 2013 12th International Conference on* (Vol. 2, pp. 54–59). IEEE.

11. Kesler, B. (2011). The vulnerability of nuclear facilities to cyber attack. *Strategic Insights, 10*(1), 15–25.

12. Lee, R. M., Assante, M. J., & Conway, T. (2016). Analysis of the cyber attack on the Ukrainian power grid. *SANS Industrial Control Systems*.

13. NUREG/CR-6882, (2006). Assessment of wireless technologies and their application at nuclear facilities. ORNL/TM-2004/317.

14. Song, J. G., Lee, J. W., Lee, C. K., Kwon, K. C., & Lee, D. Y. (2012). A cyber security risk assessment for the design of I&C systems in nuclear power plants. *Nuclear Engineering and Technology, 44*(8), 919–928.

15. Goel, S., Hong, Y., Papakonstantinou, V., & Kloza, D. (2015). *Smart grid security*. London: Springer London.

16. Mo, Y., Kim, T. H. J., Brancik, K., Dickinson, D., Lee, H., Perrig, A., & Sinopoli, B. (2012). Cyber–physical security of a smart grid infrastructure. *Proceedings of the IEEE, 100*(1), 195–209.

17. Lu, Z., Lu, X., Wang, W., & Wang, C. (2010, October). Review and evaluation of security threats on the communication networks in the smart grid. In *Military Communications Conference, 2010-MILCOM 2010* (pp. 1830–1835). IEEE.

18. Dondossola, G., Szanto, J., Masera, M., & Nai Fovino, I. (2008). Effects of intentional threats to power substation control systems. *International Journal of Critical Infrastructures, 4*(1-2), 129–143.

19. Taylor, C., Krings, A., & Alves-Foss, J. (2002, November). Risk analysis and probabilistic survivability assessment (RAPSA): An assessment approach for power substation hardening. In *Proc. ACM Workshop on Scientific Aspects of Cyber Terrorism,(SACT), Washington DC* (Vol. 64).

20. Ward, S., O'Brien, J., Beresh, B., Benmouyal, G., Holstein, D., Tengdin, J.T., Fodero, K., Simon, M., Carden, M., Yalla, M.V. and Tibbals, T., 2007, June. Cyber Security Issues for Protective Relays; C1 Working Group Members of Power System Relaying Committee. In *Power Engineering Society General Meeting, 2007. IEEE* (pp. 1–8). IEEE.

21. Alamaniotis, M., Chatzidakis, S., & Tsoukalas, L.H., "Monthly Load Forecasting Using Gaussian Process Regression," *9th Mediterranean Conference on Power Generation, Transmission, Distribution, and Energy Conversion: MEDPOWER 2014*, November 2014, Athens, Greece, pp. 1–7.

22. Qiu, M., Gao, W., Chen, M., Niu, J. W., & Zhang, L. (2011). Energy efficient security algorithm for power grid wide area monitoring system. *IEEE Transactions on Smart Grid, 2*(4), 715–723.

23. Bishop, C.M. *Pattern Recognition and Machine Learning,* New York: Springer, 2006.

24. Alamaniotis, M., Ikonomopoulos, A., & Tsoukalas, L.H., "Probabilistic Kernel Approach to Online Monitoring of Nuclear Power Plants," *Nuclear Technology*, American Nuclear Society, vol. 177 (1), January 2012, pp. 132–144.

25. C.E. Rasmussen, and C.K.I. Williams, *Gaussian Processes for Machine Learning,* Cambridge, MA: MIT Press, 2006

26. D.J.C. Mackay, Introduction to Gaussian Processes, in C. M. Bishop, editor, *Neural Networks and Machine Learning,* Berlin: Springer-Verlag, 1998, vol. 168, pp. 133–155.

27. Tsoukalas, L.H., and R.E. Uhrig, *Fuzzy and Neural Approaches in Engineering*, Wiley and Sons, New York, 1997.

28. Alamaniotis, M, & Agarwal, V., "Fuzzy Integration of Support Vector Regressor Models for Anticipatory Control of Complex Energy Systems," *International Journal of Monitoring and Surveillance Technologies Research,* IGI Global Publications, vol. 2(2), April-June 2014, pp. 26–40.

29. Consortium for Intelligent Management of Electric Power Grid (CIMEG), http://www.cimeg.com

30. Alamaniotis, M., & Tsoukalas, L., "Layered based Approach to Virtual Storage for Smart Power Systems," in *Proceedings of the 4th International Conference on Information, Intelligence, Systems and Applications,* Piraeus, Greece, July 2013, pp. 22–27.

31. Alamaniotis, M., Tsoukalas, L. H., & Bourbakis, N. (2014, July). Virtual cost approach: electricity consumption scheduling for smart grids/cities in price-directed electricity markets. In *Information, Intelligence, Systems and Applications, IISA 2014, The 5th International Conference on* (pp. 38–43). IEEE.

32. Tsoukalas, L. H., & Gao, R. (2008, April). From smart grids to an energy internet: Assumptions, architectures and requirements. In *Electric Utility Deregulation and Restructuring and Power Technologies, 2008. DRPT 2008. Third International Conference on* (pp. 94–98). IEEE.

33. Tsoukalas, L. H., & Gao, R. (2008, August). Inventing energy internet The role of anticipation in human-centered energy distribution and utilization. In *SICE Annual Conference, 2008* (pp. 399–403). IEEE.

34. Alamaniotis, M., & Tsoukalas, L. H. (2016, June). Multi-kernel anticipatory approach to intelligent control with application to load management of electrical appliances. In *Control and Automation (MED), 2016 24th Mediterranean Conference on* (pp. 1290–1295). IEEE.

Miltiadis "Miltos" Alamaniotis is a research assistant professor in the School of Nuclear Engineering at Purdue University since September 2014. He received his Dipl-Ing. in Electrical and Computer Engineering from University of Thessaly, Greece in 2005 and his M.S. and Ph.D. degrees in Nuclear Engineering from Purdue University in 2010 and 2012 respectively. His interdisciplinary research focuses on development of intelligent systems and machine learning approaches for smart power systems, smart cities and grids, security, radiation detection, and nuclear power plant controls. He had held a guest appointment with Argonne National Laboratory from 2010 to 2012, and was a visiting scientist in the Power and Energy Division of Oak Ridge National Laboratory in May 2016. In Fall 2015, he was honored with the "Paul C. Zmola Scholar" Award by Purdue University for his research contributions. He is an active member of the American Nuclear Society, IEEE, and has served as chair in artificial intelligence conferences.

Lefteri H. Tsoukalas is professor and former head of the School of Nuclear Engineering at Purdue University and has held faculty appointments at the University of Tennessee, Aristotle University, Hellenic University and the University of Thessaly. He has three decades of experience in smart instrumentation and control techniques with over 200 peer-reviewed research publications including the textbook *"Fuzzy and Neural Approaches in Engineering"* (Wiley, 1997). He directs the Applied Intelligent Systems Laboratory, which pioneered research in the intelligent management of the electric power grid through price-directed, demand-side management approaches, and anticipatory algorithms not constrained by a fixed future horizon but where the output of predictive models is used over a range of possible futures for model selection and modification through machine learning. Dr. Tsoukalas is a Fellow of the American Nuclear Society. In 2009 he was recognized by the Humboldt Prize, Germany's highest honor for international scientists.

Visualization and Data Provenance Trends in Decision Support for Cybersecurity

Jeffery Garae and Ryan K.L. Ko

Abstract The vast amount of data collected daily from logging mechanisms on web and mobile applications lack effective analytic approaches to provide insights for cybersecurity. Current analytical time taken to identify zero-day attacks and respond with a patch or detection mechanism is unmeasurable. This is a current challenge and struggle for cybersecurity researchers. User- and data provenance-centric approaches are the growing trend in aiding defensive and offensive decisions on cyber-attacks. In this chapter we introduce (1) our Security Visualization Standard (SCeeL-VisT); (2) the Security Visualization Effectiveness Measurement (SvEm) Theory; (3) the concept of Data Provenance as a Security Visualization Service (DPaaSVS); and (4) highlight growing trends of using data provenance methodologies and security visualization methods to aid data analytics and decision support for cyber security. Security visualization showing provenance from a spectrum of data samples on an attack helps researchers to reconstruct the attack from source to destination. This helps identify possible attack patterns and behaviors which results in the creation of effective detection mechanisms and cyber-attacks.

1 Introduction

Data analytics and decision support for cybersecurity through methodologies, techniques, and applications has been the core ingredient to drawing conclusions and making better judgements on cyber-attacks. Network mapping and logging techniques in applications such as Wireshark has always assisted security network analysts to understand malicious IP packets [1, 2]. Other anomaly detection systems and techniques have enabled security experts to understand threats and attacks in a deeper way through identifying attack patterns and behaviors [3, 4]. This in return helps security experts and researchers to draw better cyber security decision support conclusions.

J. Garae (✉) • R.K.L. Ko
Cyber Security Lab, Department of Computer Science, University of Waikato,
Hamilton, New Zealand
e-mail: jg147@students.waikato.ac.nz; ryan@waikato.ac.nz

© Springer International Publishing AG 2017   243
I. Palomares et al. (eds.), *Data Analytics and Decision Support for Cybersecurity*,
Data Analytics, DOI 10.1007/978-3-319-59439-2_9

While network logging and analytical methods help data analytics, data collected from modern threats and attacks are growing rapidly and new malicious attacks are more sophisticated. This requires better security approaches, methods and solutions to help understand them. Data provenance and security visualization are the growing trend for cyber security solutions [5–7]. Data captured from desired cyber-attacks creates the ability to reconstruct malicious attack landscapes and attribution to the attack origin. In addition the ability to track files end-to-end, from creation till deletion provides better decision support to cyber security experts and researchers [6, 7]. This helps security experts to identify malicious patterns and behaviors, as a result of which better conclusions are drawn and effective security implementation are taken such as security patches and defensive measures. Unlike the "ILoveYou" worm in 2000, hackers and malware developers are getting smarter by implementing cyber-attacks not only for fame or revenge, but also as targeted attacks in forms of phishing attacks, organized cybercrime attacks and nation state attacks [8]. For example, the STUXNET attack (zero day attack) on the Natanz uranium enrichment plant in Iran was regarded as the first digital weapon [9, 10]. Another example is the Sony Pictures data leak which has believed to be instigated by a nation state [11, 12]. Such cyber-threats require urgent and intelligent security methods to help protect systems and networks. However, if the attacks have penetrated systems and networks, then the need for intelligent data analytics are highly favorable. Security Visualization as a method in data analytics is a go to technique where security experts not only investigate the cyber-attacks using visualization but they also visually interpret how the attacks occur therefore reducing the analysis time spent on analyzing attack datasets [13–16]. This in return provides better decision support for the entire cyber security realm.

1.1 Motivation for This Chapter

Cybercrime has relived in a new sophisticated manner. Modern technologies have revolved around stronger security, and as technologies evolve defensively and securely. As a result, hackers and cybercriminals are stepping up their game by modifying their attack methods using smarter, sophisticated and stealth cyber-threats. Zero-day attacks are stressing the capacity for decision support technology techniques. Current analytical time on identifying zero-day attacks is unmeasurable before a patch and detection mechanism is released [17, 22]. Modern security approach and methods of countering cyber-threats and attacks have fallen short whereby users, businesses and even nations are frequently becoming targets and victims of cyber-attacks. A burgeoning attack type with the use of ransomware such as Locky ransomware is currently a new digital blackmailing attack method [18]. It has proven to be a major cyber-threat to individual users, businesses particularly in health and medical industries [19]. The Sony Pictures data breach cyber-attack has resulted in millions of losses [17, 18, 20, 21]. These form of attacks require knowledgeable security experts and most importantly security techniques, methods and applications that draw effective conclusions aiding decision support for cyber security. For example, effective security technologies that have tracking, monitoring and reporting capabilities. There is a need to harmonize threat intelligence technologies with time-base provenance technologies to provide effective and precise findings.

The focus of this chapter is on data provenance; security visualization techniques, effective visualization measurement techniques, cyber security standards and application to aid decision support for cyber security. Although data provenance is defined in many ways depending on its niche area of interest, in this chapter, data provenance is defined as series of chronicles and the derivation history of data on meta-data [22, 23]. The ability to track data from the state of creation to deletion and reconstruct its provenance to explore prominent cyber-attack patterns at any given time is the prime approach for this chapter. This is done using data collected from logging applications and security visualization [6, 7, 22].

This chapter elaborates on incorporating security visualization and data provenance mechanisms aiding data analytics and decision support for cybersecurity. The benefits of incorporating security visualization and data provenance mechanisms are as follows:

- It shares precise insights drawn from visually analyzing collected data (systems, networks and web data).
- It also provides a comparison between existing cyber security standards and establishes a new security visualization standard to aid users.

Several use cases from Tableau and Linkurious visualization platforms are used in this chapter in Sect. 4.1. In addition, we will be emphasizing more on the inclusion of security visualization into the law enforcement domain. We provide an overview of our new security visualization standard and further discuss the benefits of threat intelligence tools. And finally security visualization provides a full-proof user-centered reporting methodology for all level of audiences (CEOs, management and ordinary employees).

This chapter is organized as follows: Sect. 2 offers a background knowledge on cyber security technologies; Sect. 3 identifies common areas where cyber security technologies exists; Sect. 4 provides how cybersecurity technologies contribute to 'Decision Support' for Cyber Security; Sect. 5 provides our main contribution of research work which is the establishment of a new 'Security Visualization Standard'; Sect. 6 proposed our 'Security Visualization Effectiveness Measurement (SvEm)' theory; the concept of providing 'Data Provenance as a Security Visualization Service (DPaaSVS)' and User-centric Security Visualization; and Sect. 7 provides the concluding remarks for this chapter.

2 Background

In general, data analytics for cyber security is widely used for exploring and reporting, particularly when analyzing threat landscapes, vulnerabilities, malware and implementing better detection mechanisms. Situation awareness is a prime reporting feature in data analytics. Knowing the exact types of cyber-threat threatening organizational and banking industries is an important key indicator to implementing better security solutions [24]. Understanding different threat vectors targeting different domains, helps researchers and industries to develop specific defensive solutions.

However, business intelligence, Big data analytics provided in cloud technologies are some of the methods used to provide better understanding of the security processes in organizations. Sections 2.1 and 2.2 will share insights on current uses, trends and challenges for business intelligence in cloud technologies.

2.1 Business Intelligence and Analytics

Data analytics is widely used alongside business intelligence (BI&A) and in big data analytics [24, 25]. It is an important area of study by both researchers and industries with intentions of exploring and reporting data-related problems and to find solutions. As the Internet grows, there is an exponential increase in the type and frequency of cyber-attacks [27]. Sources ranging from data warehouses to video streaming and tweets generate huge amount of complex digital data. Cloud technologies provide scalable solutions for big data analytics with efficient means of information sharing, storage and smart applications for data processing [26, 28]. Gartner estimated that by 2016, more than 50% of large companies data will be stored in the cloud [27, 29]. Big data analytics using data mining algorithms that require powerful processing power and huge storage space are an increasingly common trend. It has reporting mechanism and often visual dashboards. However, because CEOs and upper-level managers are not always tech-savvy, lack of clarity and complexity with information acquired, makes the comprehensive reporting on such analytics a difficult task for cyber security experts. This is a challenge which often raises concerns in decision making situations. For example, data breach magnitude and the assessment process are often under estimated, not not reported clearly. This affects how mitigation processes to resolve the situation. As a result, the organization's reputation can be at stake and such organizations are vulnerable to cyber-attacks.

2.2 Big Data and Cloud Technologies

A major application in big data analytics is parallel and distributed systems. This method coexists as part of the entire cloud infrastructure to sustain exceeding exabytes of data and the rapid increase rate in data size [30]. The need to frequently increase processing power and storage volumes are critical the factor for cloud infrastructures. Adding onto this, security, fault-tolerance and access control are critical for many applications [30]. Continuous security techniques are built to maintain these systems. This is yet another key decision factor in organizations and industries for cyber security frameworks. Cloud technologies also provide scalable software platforms for the use of smart grid cyber-physical systems [31]. These platforms provide adaptive information assimilation channels for ingesting dynamic data; a secure data repository for industries and researchers [31, 32]. It is a trend for power and energy companies whereby data has become valuable and further investigations into the data for insights are required and relevant for better service delivery. While information sharing and visual data analytics are useful features in these systems, data security is still major concern. With current sophisticated cyber-attacks involving phishing or social engineering elements, customer data and utility data are the main target [33, 34].

3 Cyber Security Technologies

The technological shift and drift from common use of desktop computers to mobile platforms and cloud technologies have expanded the cyber-threat landscapes [45–47]. Newer urgent needs emerged in as the digital economy has matured over the past decade. Businesses and consumers are more dependent than ever on information systems [48]. This has contributed to how cyber security has evolved over the past decade. The cyber security focus in the past decade for researchers and industries can be summed up with the list below [45–48]:

1. *Endpoint Detection and Response:* These technologies includes intrusion detection systems [36], provide the ability to frequently analyze the network and identify systems or applications that might be compromised. With endpoint detection mechanisms, responsive steps can be taken to mitigate cyber-attacks [35].
2. *Network and Sensors for Vulnerabilities:* Such technologies provide flexibilities to both users and network operators. Either wireless or wired, the sensor network has multiple preset funtions such as sensing and processing, to enable multiple application goals [49]. Senor nodes are capable of monitoring the network area with the aim identifying and detecting interested security events and reporting to a base station deployed on the network.
3. *Cloud Security:* The emerging cloud technologies have offered a wide range of services including network, servers, storage, range of applications and services [37]. However, this brought in a new range of security challenges which have contribute to how cloud technologies have transformed from the past 10 years.
4. *Cyber Threat Intelligence:* Cyber threat intelligence technologies profiles a holistic approach for automated sharing, real-time monitoring, intelligence gathering and data analytics [38]. Organizations are emphasizing on cyber threat intelligence capabilities and information sharing infrastructure to enable communications and trading between partners [39].

5. *Hardware & Software Security:* These includes providing security for hardware products and software products. Current trends indicated that hardware and software technologies have added capabilities and functions which require security components to safeguard networks and applications.

6. *Security Tools:* Security tools generally covers applications which are often used for securing systems and networks, e.g. penetration testing tools, vulnerability scanners and antivirus softwares.

7. *Security Tracking and Digital Auditing:* These technologies focus on auditing purposes especially to observe and record changes in the systems and networks. Configuration changes of a computerized device by tracking the processes and system tracks modification are some examples [40]. Other security tracking purposes include geo-location tracking, monitoring operational variables and outputs of specific devices of interest [44].

8. *User and Behavioral Analytics:* These technologies emphasize on understanding user behaviors, behavioral profiles and end users. Security concerns over targeting inappropriate audience are some of the issues encountered with these technologies [41].

9. *Context-aware Behavioral Analytics:* Context-aware technologies provide application mechanisms that are able to adapt to changing contexts and able to modify its behavior to suit the user's needs, e.g. smart homes inbuilt with a context aware application that can alert a hospital if a person urgently requires medical attention [42].

10. *Fraud Detection Services:* As Fraud cases are becoming popular, computer based systems designed to alert financial institutions based on set fraud conditions used to analyze card-holder debits. These systems also identify 'at risk' cards which are possessed by criminals [43].

11. *Cyber Situational Awareness:* Recent research and surveys have shown that the rise of cyber criminal activities have triggered the need to implement situational awareness technologies. These includes the use of surveys, data analytic and visualization technologies.

12. *Risk Management Analytics:* Risk management analytics are methodologies used by organizations as part of their business strategies to measure how well their business can handle risks. These technologies also allows organizations to predict better approach to mitigate risks.

The research and industry interests are targeting mainly end-users, data collected, security tools, threat intelligence and behavioural analytics [46–48, 50]. For example in context-aware behavioral analytics technologies, we can witness techniques such as mobile location tracking, behavioral profiling, third-party Big data, external threat Intelligence and bio-printing technologies. These are founded on the principles of unusual behavior or nefarious events [50]. Overall, popular cyber security technologies are summarized in the following categories as shown in Fig. 1 [50]. Honeypots are examples of active defense measures. Cloud-based applications and BYODs technologies are far beyond the realm of traditional security and firewall. They are well suited for the cloud and Security Assertion Markup Languages (SAML)

Fig. 1 Cyber security technologies trend overview in 2016

combined with intrusion detection technologies and encryption to maintain control in corporate networks [51, 52]. Early warning systems using smart algorithms are able to determine and alert security professionals on which sites and servers are vulnerable to being hacked [50].

4 Decision Support for Cyber Security

In Cyber Security, the notion of Decision Support or decision support technologies are of prime importance for corporate businesses. Early warning systems, risk management analytics, security tracking and digital auditing systems are giving corporate businesses and researchers the ability to make better decisions. Traditional decision support technologies heavily rely on data analytic algorithms in order to make important security decisions on cyber-attack and data breaches. This is often based on post-data analytics to provide reports on attack patterns and behaviors. Cyber security countermeasures as part of risk planning are effective to ensure

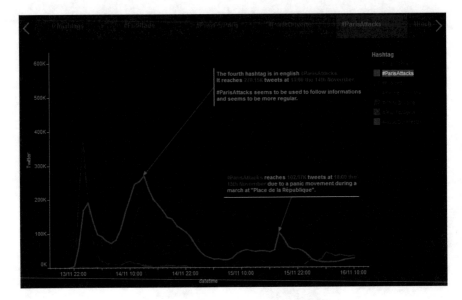

Fig. 2 Tableau visualization—analysis on twitter Hashtags on the Paris attacks

confidentiality, availability and integrity of any information system [53]. Decision support for cyber security is critical for researchers but especially for corporate organizations. It affects the organizations reputation particularly its daily operations. The ability to minimize the uncertainties in cyber-threat rates, countermeasures and impacts on the organization resulting in a low cybersecurity costs [53].

4.1 Visual Analytics for Cyber Security

In this subsection, we assess a number of leading cyber security visualization platforms and discuss the directions targeted by each of these platforms with regards to aiding decision making processes. These assessments are based on existing use cases.

Current data analytics applications have enhanced reporting features by including visualization as part of their reporting templates. For example, the use of visual dashboards is highly popular with impressive statistical analytic benefits. Most business intelligence & analytics (BI&A) systems use visual dashboards to provide statistical reports. They displays a multidimensional information sharing platform with minimal space required to display findings and has changed the way reporting is presented in the twenty-first century. BubbleNet is a visualization dashboard for cybersecurity which aims to show visual patterns [55] and Tableau is a visualization dashboard tool that offers live data visualization by connecting to local or remote data [56]. For example, the Paris attack twitter communication as shown in Fig. 2 was largely spread by social media and Edward Snowden's Twitter analysis as

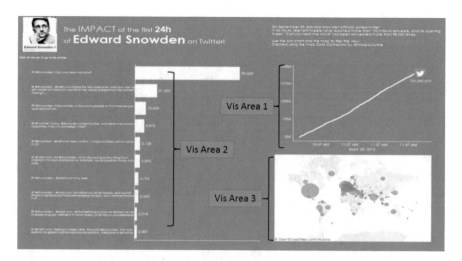

Fig. 3 Tableau visualization dashboard of Edward Snowden's impact on Twitter

shown in Fig. 3. Tableau has capabilities of loading raw data and dashboard display columns that could accommodate different visual reports of a given data collected. This is illustrated with "Vis Area 1, 2, and 3" labels in Fig. 3.

Linkurious, a web-based graph data visualization platform has capabilities to visualize complex data from security logs (network, system, application logs) [56]. It is a graph data visualization for cyber security threat analysis with the aims of (1) aiding decision support and (2) providing intelligence to support the decisions made [63]. Below are several Linkurious threat visualization analysis shown in Figs. 4, 5 and 6.

Figure 4 shows a Linkurious visual analytics of a "normal network behaviour." With such known visual patterns of what a normal network behaviour is, any future changes in the visual pattern will prompt security experts to further investigate it. For example, comparing a visual analytics pattern in Fig. 5 (a UDP Storm Attack) with Fig. 4, Cyber Security experts can effectively spot the difference and conclude that Fig. 5 shows a possibly malicious network traffic. A storm of incoming network packets targeting the IP: 172.16.0.11, as indicated by the direction of the arrows on the visualization clearly indicates that this is a denial of service attack (DoS). Linkurious also has the capabilities of representing data relationships between entities. In Fig. 6, visual representations of IP addresses are shown indicating data packets movement from the source IP address to the destination IP address.

Pentaho, a business analytics visualization platform, visualizes data across multiple dimensions with the aim of minimizing dependence on IT [58]. It has user-centric capabilities of drill through, lasso filtering, zooming and attribute highlighting to enhance user experiences with reporting features. It provides user interactions with the aim of allowing users to explore and analyze the desired dataset.

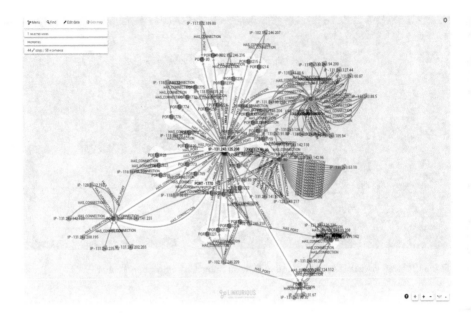

Fig. 4 Linkurious visualization threat analysis—normal network behaviour

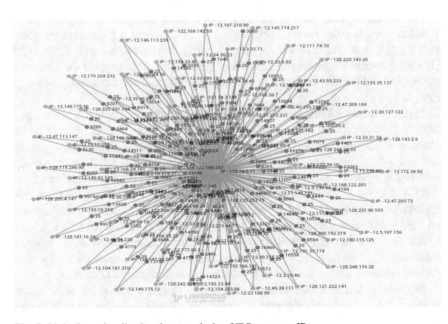

Fig. 5 Linkurious visualization threat analysis—UDP storm on IP

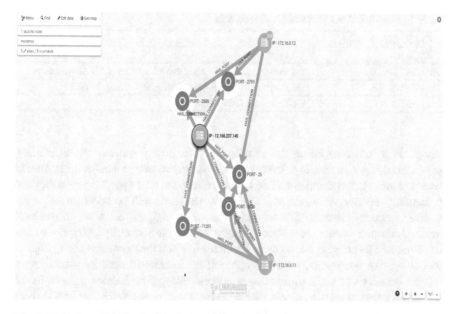

Fig. 6 Linkurious visualization IP relationship threat analysis

The ability to include visualization in these applications provides better reporting capabilities for security experts and researchers. OpenDNS with its OpenGraphiti visualization platform has features pairing up visualization and Big Data to create 3D representations of threats [54]. This enables visual outputs which practically do not require much additional explanation, due to the simplicity and self-explanatory visual outputs. Linkurious provides Intelligence analysis, fraud detection mechanisms and cyber security attack analysis mechanisms [57]. Other security companies such as Kaspersky, Norse, and Fireeye also provide real-time visualization platforms to help security researchers to better understand cyber-attacks and therefore make better decisions [59–62].

5 The Need: Security Visualization Standard

Although there is a vast range of cyber security tools, and software applications publicly available, there is still a need for data analytics and decision support mechanisms targeting specific cyber security landscapes. This is because, users (CEOs, Management and analysts) may belong to different security background and exhibit different levels of knowledge. Specific user-centric cyber security tools and techniques targeting specific audiences requires some specific cyber security standard. Cyber security decision support standards are important to guide and habour all reporting processes. Security researchers and experts leverage on such

Table 1 Standards directly related to cyber security

ISO/IEC 27000 series	
ISO Codes	Description
ISO/IEC 27000	Information Security Management Systems (ISMS)—overview
ISO/IEC 27001	Information Technology: Security Techniques in ISMS
ISO/IEC 27032	Guideline for Cybersecurity

standards in order to be on the same page with how a security event is being presented using visualization. This means creating a security visualization standard would provide a scope and guideline to help reduce the time spent on producing and presenting security visualizations. Therefore, less time will be spent on assessing a given security visualization to gain the most insights on a security visual event. Although current academic research centers and security enterprises have implemented in-house cyber security standards that suit their research aims, projects and enterprise operations, the challenge and complexity in security visualization reports manifests to some difficulties in understanding visualization reports. In the following subsection, we present various cyber security standards and assess them with the purpose of highlighting the importance how cyber security standards help manage and reduce possible cyber-attacks.

5.1   Outline on Cyber Security Standards

As part of the International Organization for Standarization/International Electrotechnical Commission 27000 series, 27001 (ISO/IEC 27001:2013)—Information security management and ISO 27032—Guideline for cybersecurity, has established guidelines and certification standard for Information security and cybersecurity as a whole [68–70]. Table 1 summarizes the cyber security standards. From a practical perspective, ISO 27001 formally provides a guideline of managing information risks through information security controls in an organization. For example, the ISO/IEC 27001:2005 provided the "PDCA (Plan-Do-Check-Act/Adjust) cycle or Deming cycle." ISO 27032 on the other hand is a cybersecurity certification standard on Internet Security [70]. In any security standards, policies and guidelines are implemented to include and maintain all aspects in security events. Alternatively, the use of visualization in cyber security should have a Security Visualization Standard [71, 72].

Table 2 Types of cyber security visualization guidelines

Cyber security visualization guidelines	
Guidelines	Description
Representation guidelines	How information is represented
Selection guidelines	Visualization platforms
Color guidelines	Guidelines on how to select and use colors
Shape guidelines	Guidelines on choices of shapes used
User guidelines	User-interaction guidelines

5.2 SCeeL-VisT: Security Visualization Standard

Although different organizations have in-house visualization standards according to their preferences, interpretation complexities from visualization platforms to users without a solid technical background is a common concern in the cyber security domain. Interpretation accuracy with meaningful and useful insights in a time sensitive manner is a key feature when using security visualization.

In most cyber security research centers, particularly in law enforcement, security use cases are presented with research outputs often with sensitive data that require urgent need to safeguard and preserve them with policies, physical secure storage and guidelines. These policies and guidelines outlines how to handle and use these sensitive data. Security visualization for law enforcement requires a security visualization standard to maintain information security, information sharing and exchanging of insights presented from visualizing of security events. This standard will act as basis to implementing and presenting security visualization with the aim to maintain the confidentiality and integrity of the underlying raw data. We present our Security Visualization Standard (SCeeL-VisT) model with related descriptions. Further detail on SCeeL-VisT standard is given in the following pages. The SCeeL-VisT standard has two major parts:

1. Understanding what is in Hand (Dataset)
2. Security Visualization Process

5.3 Security Visualization Policies and Guidelines

On a broader coverage of the Standard, a way forward for cyber security researchers and industry security experts is to establish security visualization policies and guidelines. These guidelines are shown in Table 2.

The purpose of developing SCeeL-VisT standard is to put emphasis on common grounds with security visualization development, presentation, and understanding security events. Because the use of data visualization has been widely used across many research domains and industries, complying with a security visualization standard is of vital importance. This is due to the high frequency in which sensitive data must be dealt with. In addition to that, current security visualizations are tailored towards specific audiences making it difficult for other interested users to understand insights presented in a given visualization. Finally, the most crucial purpose of having a standard like SCeeL-VisT is its contribution to making effective decisions in cyber security. Such standards creates a clear and precise scope for both cyber security experts and what security visualizations should show in relations to the raw data used.

5.4 Law Enforcement Visualization Environment

Cyber security for law enforcement organizations especially international law enforcement agencies are seeing new trends emerging in highly complex cybercrime networks. Generally, law enforcement agencies such as Interpol have classified "Cybercrime" in two main types of internet-related crime [64]:

- **Type 1: Advanced cybercrime** (or high-tech crime)—Sophisticated Attacks against computer hardware and software.
- **Type 2: Cyber-enabled crime**—Many 'traditional' crimes have taken a new turn with the advent of the Internet, such as crimes against children, financial crimes and even terrorism.

For the law enforcement environment, cybersecurity technologies are driven towards aiding investigations, information sharing, securing and preserving data. Examples of this include malware threat intelligence, defensive and mitigation technologies [57, 59–62]. Cyber security today is driven by the way cybercriminals are executing cyber-attacks. Most cyber-attacks are increasingly smart and sophisticated coercively drives law enforcement, researchers and industry experts to match up with both defensive and offensive security applications. Data analytics and threat intelligence are some of some of the examples used for tracking and monitoring interested cyber criminal entities. As the cybercrime nature grows from individual and small group cybercriminals to advance methodologies, we see highly complex cyber criminal networks from both individuals and organized cyber criminal organizations. Current cyber attacks stretch across transnational jurisdictions in real-time to execute cyber-attacks on an unprecedented scale. Interestingly, the flow of digital currencies movement are associated with these cybercrimes, especially organized cybercrimes. Therefore the cyber security trend for law enforcement can be scaled down to three main technology categories: (1) Attribution tools; (2) Cyber Threat Intelligence and (3) Secure Information Sharing Technologies (SIST). However, the rate at which malwares and ransomwares are created versus the implementation of new cybersecurity tools to combat these cyber-threats is far beyond proportion. This raises a concern for law enforcement and cyber security across international borders.

Existing cybercrime challenges for law enforcement range from malware attacks, ransomwares and terrorism [64]. Security approaches such as data analytics and threat intelligence methodologies are some of the steps taken by law enforcement research domains to help with improving investigation technologies. Due to how cybercrime has been broadening its attack environment from not just the technology domain but broadly penetrating others such as health and financial domain, the fight against cybercrime for law enforcement has taken extensive direction which requires external and international collaborations. This means information sharing legislations have to be re-visited, policies and guidelines have to be implemented.

The rise of dark market trading between cybercriminals on the dark web allows them to produce and escalate cyber-attacks at a rate leaving law enforcement, academia and industry researchers to encounter cyber security challenges. However, associating the use of digital currencies (Bitcoins) by cybercriminals with cyber-crimes, triggers the need for new threat intelligence tools, data analytics tools and vulnerability scanners that would allow effective execution of investigations.

Threat Intelligence tools and 24/7 monitoring live-feed applications allow law enforcement agencies to carry out their investigations effectively especially for transnational cybercrimes whereby international collaborations is required. Information sharing between international law enforcement agencies without sharing the underlying raw data capabilities is of high demands.

Bitcoin[1] [65] is a peer-to-peer distributed electronic cash system that utilizes the blockchain[2] protocol [66, 67]. Due to how bitcoin payment transactions operate over the Internet in a decentralized trustless system, law enforcement agencies are seeking ways aid their investigations especially to track and monitor Bitcoins movements that are involved in cybercrime events. The ability to provide visual threat intelligence on Bitcoin transactions using blockchain tools are some of the way forward to fighting cybercrime. Where the money flows to and from known cybercriminals, allows law enforcement and cyber security investigators to track cybercrime events.

---

[1]Bitcoin is a distributed, decentralized crypto-currency, which implicitly defined and implemented Nakamoto consensus.

[2]Blockchain is a public ledger of all Bitcoin transactions that have executed in a linear and chronological order.

Security Visualization Standard (SCeeL-VisT)

Part 1: Understanding what is in Hand (Dataset)

1. **The Problem**:- Identification of Security Event

   - Identify Security Event (e.g. Malware Attack, SQL Injection, etc.) & Data Type (Raw data: log files, Social media data, etc.)
   - Understand the nature of data (e.g. financial data, Health data. etc.)

2. **The Purpose**: Know the Visualization Type and Technique

   - Understand the intension of security Visualization (e.g. show relationships, etc.)
   - Decision: Exploratory or Reporting Visualization
   - Decision: Security visualization Technique (e.g. Categorizing: time-base, Provenance-base attack-base)

3. **Cyber Attack Landscape**: Know the cyber-attack location (e.g. Network, Systems, Application layer, etc.)

   - Know the point of attack (e.g. network attack, identify source and destination of attack, etc.)
   - Attribution of cyber-attack

Part 2: Security Visualization Process

1. **Visual Presentation Methodologies**: How to present data visually
2. **Color & Shape Standard for Security**: Decision on choice of colors

   - Standardizing Main Color choices

     - Color: Red = High Attack Nature or Violation (e.g. Malware Process)
     - Color: Yellow = Suspicious Process (e.g. IP address)
     - Color: Green = Good or Normal Process (e.g. Network traffic)
     - Color: Blue = Informational Process (IP address)
     - Color: Black = Deleted, Traces: non-existed (e.g. deleted file,)

   - Standardizing Main Shapes choices

     - Shape: Circle = Nodes (e.g. Network Nodes)
     - Shape: Rectangle = Files (e.g. .docs, .jpeg, etc.)
     - Shape: Square = Data Clusters (e.g. IP address—network traffic)
     - Shape: Diamond = Web / Social Media Process (Social media data)

(continued)

- Standardizing Use of Line Types
  - Line: Single Line (—) = Relationships, Connections, Links, Provenance, time-base (e.g. between two Network Nodes)
  - Line: dotted Line (- - -) = Possible Relationships (e.g. .docs, .jpeg)
  - Line: Solid Arrow (→) = Direction of relationship or interaction
  - Line: Dotted Arrow (— >) = Predicted Relationship or Interaction

3. **Security Visualization Techniques**: Provenance & Attribution-base, User-centered, Real-time base
4. **Security Visualization Type**: Animated, 3D, Static, etc.

6 Effective Visualization Measurement Techniques

Prior to this section, cyber security technology trends and use cases were analyzed and discussed including security visualization. These are part of the growing trend and effective methodologies for addressing cyber security issues (cybercrime). However, to improve security visualization technologies, we have to look into 'how' and 'what' makes security visualization appealing to viewers of the visualization. Measuring how effective, useful and appropriate security visualizations are to users helps provide researchers and developers ways of improving security visualization technologies. Before we discuss our new 'effective visualization measurement technique', a summary of existing effectiveness measurement techniques are outlined in Table 3.

6.1 The Security Visualization Effectiveness Measurement (SvEm) Theory

By analyzing existing techniques and understanding how they function, we introduce our new effective measurement technique which measures the effectiveness in both a given security visualization and the viewer's experiences on security events. It is based on both the visualization approach and user intuition. We refer to this new effective measurement technique as 'Security Visualization Effectiveness Measurement (SvEm) theory and is displayed in Eq. (1). Elaborating on thetheorem,

Table 3 Existing visualization measurement techniques

Summary—visualization effectiveness measurement techniques	
Effective measurement factor	Range of measurement (quantity)
Cognitive load	High (germane/intrinsic/extraneous cognitive load)
Working memory (prior knowledge)	High (effects to cognitive load)
NASA-TLX test (indirect-work load)	Medium (based on work load)
Subjective Workload Assessment Technique (SWAT)	Medium (scale rating factors—mental effort)
Image quality assessment	Medium—high (based on distortion)
Eye tracking (CODE theory-visual attention)	High (eye movement based—information theory)
Brain activity	High
Response time on task (s)	Low (depends on prior knowledge and effort)
Effort/difficulty rating	Low (based on insights)
User interactions/performance	Low (based on naive physics; body, social, environmental awareness skills)
Visual perception metrics (visualization efficiency)	Low (based on graphical methods e.g. similarities)

the SvEm[3] theory aims to minimize the time (duration) spent on viewing a visualization and making sense out from the intended insights portrayed in a visualization. The components of the SvEm theorem are:

1. *Mobile platform screen surface area (w * h)*: This refers to the surface area used to display a security visualization. Screen sizes have a great impact on how visualizations appear.
2. *Security Visual Nodes (Sv_f)*: These are known security attributes identified in a visualization, e.g. an malicious or infected IP address.
3. *N-dimensions*: N-dimensions refers to how many visual dimensions used to represent the visualization. The higher number of dimensions are used for visualization indicates the depth of data being able to represent visually.
4. *User Cognitive Load (Cl)*: This is based on how much knowledge (Prior knowledge) a user has on the expect visualization. It is a prerequisite security knowledge around expected security events such as a malware cyber-attack.
5. *Memory Efficiency (t_{me})*: This is a time-base attribute which measures how fast one can recall security related attributes.
6. *Number of Clicks (n_{clicks})*: This refers to how many clicks one has to perform on the mobile platform screen in order to view the required visualization.

---

[3]The Security Visualization effectiveness Measurement theory designed for mobile platforms is measured in percent (%) provides a way to measure clarity and visibility in a given security visualization.

Security Visualization Effectiveness Measurement (SvEm) Theory

$$SvEm = \frac{(w * h)/Sv_f * d_n}{Cl * t_{me} * n_{clicks}} > 50\%(Distortion) \qquad (1)$$

Where:

w * h : Mobile Display Area (dimensions)

Sv_f : Security Visual Nodes (e.g. Infected-IP, Timestamp, etc.)

d_n : n-dimensional view of security visualization

Cl : Cognitive Load (Identifiable Attributes (Quantity)—Prior Knowledge)

t_{me} : Memory efficiency (Effort based on Working memory—Time-base)

n_{clicks} : Number of clicks on Visualization

The theoretical concept behind SvEm is based on two main effectiveness measurement factors:

1. **Visualization Distortion**: Effectiveness in Distortion is defined when a visualization has *greater than* 50% visual clarity

   - How clear and readable visualization is presented
   - Features and Attributes presented are visible
   - Visual patterns emerge into reality to observers.

2. **User Response Time**: Duration in milliseconds (ms)

   - Analytical time of processing the visualization
   - Time of user- cognition to recall known memory

Based on this theory, we have observed that the factors highly contributing to a high SvEm value are: (1) *w * h: smartphone dimensions* and (2) *d_n: n-dimensional view of security visualization* i.e. a 3-dimensional representation visualization view has proven less distorted than a single-dimensional visualization view. More data information are visible and shown in higher n-dimensional visualization views. This affects the users (viewer) ability to provide a higher count for Sv_f. The less value of n_{clicks}, indicates that the overall *time* spent on viewing the visualization. This contributes to higher effectiveness measurement outcome. However, the current focus is on the following:

- Mobile platform screen resolution types
- Users cognitive load (CL: prior knowledge)
- Data input size.

Due to these contributing challenge factors, assumptions are established as guides, to give a better effectiveness measurement reading. As a result, errors are minimized to achieve appropriate and reasonable SvEm reading.

- If a user is able to understand a given security visualization within *less than* 5 s, then visualization has effective SvEm output.
- Input data has to be within the capable processing power of mobile platform used.
- User must always have some form of cognitive ability (Prior knowledge) before engaging the given security visualization.
- Number of Clicks (n_{clicks}) refers to number of clicks(navigating) on the mobile platform screen to the point where the first known security attribute has been identified.

6.2 Data Provenance as a Security Visualization Service (DPaaSVS)

Many cyber security researchers are investing their efforts into finding technological solutions in understanding cyber-attack events, defending against them and finding ways to mitigate such cyber-attacks. A prominent method in understanding cyber-attack events is related to the concept of 'Data Provenance' which is defined as "a series of chronicles and derivation history of data on meta-data" [7, 22]. Including data provenance into security visualization allows cyber security researchers and experts to be able to analyze cyber-attacks from it's origins through to it's current state i.e. from the instant when the cyberattack was found in the systems to the state of mitigation or even further to the 'post - digital forensic' state of the cyberattack. Having the ability to apply data provenance as a security visualization service (DPaaSVS), cybersecurity experts will better understand cyber-attacks, attack landscapes and the ability to visually observe attack patterns, relationships and behaviors in a nearly real-time fashion [73]. For example, Fig. 7 presents a deep node visualization of network nodes, with patterns highlighted colors, rings-nodes and lines of nodes captured every 5 s [73]. Although provenance has been used as exploratory features in existing visualization platforms, we present the concept of 'Data Provenance as a Security Visualization Service (DPaaSVS) and its features.

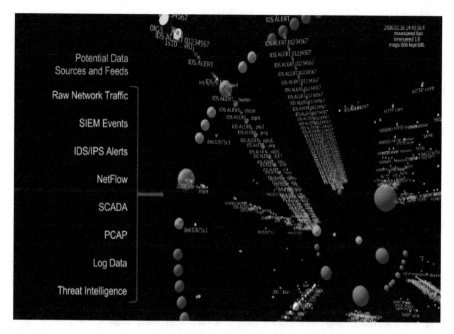

Fig. 7 Security visualization with Deep Node Inc. provenance time-base patterns

Data Provenance as a Security Visualization Service (DPaaSVS) features:
- Tracking and observing cyber-attacks from past to present using historical data.
- Threat Intelligence and monitoring.
- Educating viewers (users) with cyberattack types and attack landscapes.
- Exploring cyber-attacks using visual patterns and behaviors.
- Pinpointing cyber-attacks using timestamps.
- Empowering viewers (users) to interact with cyber security events using data collected.
- Empowering viewers (users) to effectively make decisions on cyberattack issues.

Generally, both offensive and particularly defensive security technologies depend on '*Past Knowledge*.' Security visualization based on past knowledge (Provenance) are able to show matching patterns and behaviours while observing a given visualization therefore gives insights on who such cyber-attacks are penetrating network systems. Based on provenance Cyber defenses including Firewalls, Intrusion Prevention Systems (IPS) and Antivirus are able to implement secure approaches such as:

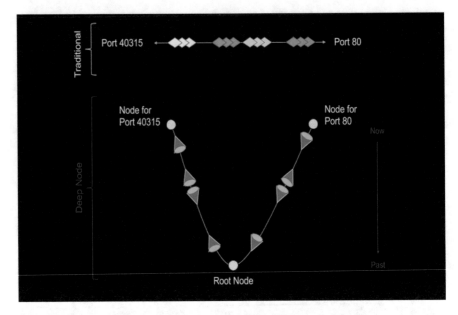

Fig. 8 Deep Node security visualization with provenance base knowledge

- *pre-defined* signature-based rules to block known suspicious traffic.
- capabilites to monitor *signs* of known malicious activities over the network traffic (web).
- capabilities to see *known* attack traffics and protocol violations over the network.

For example, Deep Node security visualization provides it's security analysts with the ability to use Data Provenance as a Security Visualization Service (DPaaSVS) by understanding network traffics based on Nodes in the concept of *Past-Node* and present [73, 74]. Nodes visualized are monitored through network traffics via it's corresponding communication *Port*. See Fig. 8.

6.3 User-Centric Security Visualization

Security visualizations like any other general visualization has a purpose. Data analytics, data exploration and reporting are the most common areas for security visualization. However, the art of presenting the visualization to the viewers (users) defines how useful it is. With targeted audiences, presenting visual insights enables users make smart and effective decisions. While there area lot of aspects to discuss under user-centric security visualization, this subsection highlights the core importance on what makes security visualization a user-centric approach and highlight important user-centric features in security visualizations. But before we look further into the details of users and their interactions with visualizations, the term 'User-centric Security Visualization' is defined in the context of this chapter. The term 'User-centric Security Visualization' refers to visualizations that empower users to interact with any given visualization in platforms to observe and explore security events. To name a few, such user-interactions features include:

- *Mouse-over click* control, labeling and tagging features.
- *Zoom-in and zoom-out* features.
- *Lasso filtering and Attribute highlighting* features
- *Multi-dimensional* (e.g. 3-Dimension (3D)) visual views.
- *Visual Animation* features.
- *Virtual reality* capabilities.
- *data input* features.
- *Color categorization* features.
- *Time and abstracts insertion* control features.

These user-interactions in visual platforms enables users to have a sense of control and interaction when visually interacting with security events based on data collected or at real-time. Such interactions naturally installs the viewers interests and instincts in motivating them to engage and understand security events in a realistic approach therefore enhances user experience and contributes to making decisions effectively.

7 Concluding Remarks

In summary, cyber security technologies are driven by Internet users and industry demands to meet security requirements to secure their systems and networks. As technologies evolve, existing cyber security technological trends are often dictated by smart sophisticated cyber-attacks causing cyber security experts to step up their game into security researches. This motivates research opportunities for data provenance and security visualization technologies in aid of understanding cyber-attacks and attack landscapes.

Due to the increasing statistics of cyber-attacks penetrating networks, existing cyber security technologies for 'decision support' are directed mainly into data analytics and threat intelligence. Understanding how to prevent, protect and defend systems and networks are the prime reasons for data analytics and threat intelligence technologies. However, Security Visualization in the context of data provenance and user-centric approaches are increasingly common and driven into the cyber security technologies for smarter and effective decision support reporting. Finally, with specific directed cyber security standards, policies and guidelines for security visualization, effective decisions and conclusions can be reached with minimal time required to react, defend and mitigate cyber-attacks.

Acknowledgements The authors wish to thank the Cyber Security Researchers of Waikato (CROW) and the Department of Computer Science of the University of Waikato. This research is supported by STRATUS (Security Technologies Returning Accountability, Trust and User-Centric Services in the Cloud) (https://stratus.org.nz), a science investment project funded by the New Zealand Ministry of Business, Innovation and Employment (MBIE). The authors would also like to thank the New Zealand and Pacific Foundation Scholarship for the continuous support towards Cyber Security postgraduate studies at the University of Waikato.

References

1. Orebaugh, Angela, Gilbert Ramirez, and Jay Beale. Wireshark & Ethereal network protocol analyzer toolkit. Syngress, 2006.
2. Wang, Shaoqiang, DongSheng Xu, and ShiLiang Yan. "Analysis and application of Wireshark in TCP/IP protocol teaching." In E-Health Networking, Digital Ecosystems and Technologies (EDT), 2010 International Conference on, vol. 2, pp. 269–272. IEEE, 2010.
3. Patcha, Animesh, and Jung-Min Park. "An overview of anomaly detection techniques: Existing solutions and latest technological trends." Computer networks 51, no. 12 (2007): 3448–3470.
4. Yan, Ye, Yi Qian, Hamid Sharif, and David Tipper. "A Survey on Cyber Security for Smart Grid Communications." IEEE Communications Surveys and tutorials 14, no. 4 (2012): 998–1010.
5. Tan, Yu Shyang, Ryan KL Ko, and Geoff Holmes. "Security and data accountability in distributed systems: A provenance survey." In High Performance Computing and Communications & 2013 IEEE International Conference on Embedded and Ubiquitous Computing (HPCC_EUC), 2013 IEEE 10th International Conference on, pp. 1571–1578. IEEE, 2013.
6. Suen, Chun Hui, Ryan KL Ko, Yu Shyang Tan, Peter Jagadpramana, and Bu Sung Lee. "S2logger: End-to-end data tracking mechanism for cloud data provenance." In Trust, Security and Privacy in Computing and Communications (TrustCom), 2013 12th IEEE International Conference on, pp. 594–602. IEEE, 2013.
7. Ko, Ryan KL, and Mark A. Will. "Progger: an efficient, Tamper-evident Kernel-space logger for cloud data provenance tracking." In Cloud Computing (CLOUD), 2014 IEEE 7th International Conference on, pp. 881–889. IEEE, 2014.
8. Bishop, Matt. "Analysis of the ILOVEYOU Worm." Internet: http://nob.cs.ucdavis.edu/classes/ecs155-2005-04/handouts/iloveyou.pdf (2000).
9. D. Kushner, The Real Story of Stuxnet, IEEE Spectrum: Technology, Engineering, and Science News, 26-Feb-2013. [Online]. Available: http://spectrum.ieee.org/telecom/security/the-real-story-of-stuxnet.
10. A. K. Z. K. Z. Security, An Unprecedented Look at Stuxnet, the Worlds First Digital Weapon, WIRED. [Online]. Available: https://www.wired.com/2014/11/countdown-to-zero-day-stuxnet/.
11. Rigby, Darrell, and Barbara Bilodeau. "Management tools & trends 2011." Bain & Company Inc (2011).
12. Bonner, Lance. "Cyber Risk: How the 2011 Sony Data Breach and the Need for Cyber Risk Insurance Policies Should Direct the Federal Response to Rising Data Breaches." Wash. UJL & Pol'y 40 (2012): 257.

13. Siadati, Hossein, Bahador Saket, and Nasir Memon. "Detecting malicious logins in enterprise networks using visualization." In Visualization for Cyber Security (VizSec), 2016 IEEE Symposium on, pp. 1–8. IEEE, 2016.
14. Gove, Robert. "V3SPA: A visual analysis, exploration, and diffing tool for SELinux and SEAndroid security policies." In Visualization for Cyber Security (VizSec), 2016 IEEE Symposium on, pp. 1–8. IEEE, 2016.
15. Rees, Loren Paul, Jason K. Deane, Terry R. Rakes, and Wade H. Baker. "Decision support for cybersecurity risk planning." Decision Support Systems 51, no. 3 (2011): 493–505.
16. Teoh, Soon Tee, Kwan-Liu Ma, and S. Felix Wu. "A visual exploration process for the analysis of internet routing data." In Proceedings of the 14th IEEE Visualization 2003 (VIS'03), p. 69. IEEE Computer Society, 2003.
17. Wang, Lingyu, Sushil Jajodia, Anoop Singhal, and Steven Noel. "k-zero day safety: Measuring the security risk of networks against unknown attacks." In European Symposium on Research in Computer Security, pp. 573–587. Springer Berlin Heidelberg, 2010.
18. Mansfield-Devine, Steve. "Ransomware: taking businesses hostage." Network Security 2016, no. 10 (2016): 8–17.
19. Sgandurra, Daniele, Luis Muñoz-González, Rabih Mohsen, and Emil C. Lupu. "Automated Dynamic Analysis of Ransomware: Benefits, Limitations and use for Detection." arXiv preprint arXiv:1609.03020 (2016).
20. Davis, Thad A., Michael Li-Ming Wong, and Nicola M. Paterson. "The Data Security Governance Conundrum: Practical Solutions and Best Practices for the Boardroom and the C-Suite." Colum. Bus. L. Rev. (2015): 613.
21. L. Widmer, The 10 Most Expensive Data Breaches | Charles Leach, 23-Jun-2015. [Online]. Available: http://leachagency.com/the-10-most-expensive-data-breaches/.
22. J. Garae, R. K. L. Ko, and S. Chaisiri, UVisP: User-centric Visualization of Data Provenance with Gestalt Principles, in 2016 IEEE Trustcom/BigDataSE/ISPA, Tianjin, China, August 23–26, 2016, 2016, pp. 1923–1930.
23. Zhang, Olive Qing, Markus Kirchberg, Ryan KL Ko, and Bu Sung Lee. "How to track your data: The case for cloud computing provenance." In Cloud Computing Technology and Science (CloudCom), 2011 IEEE Third International Conference on, pp. 446–453. IEEE, 2011.
24. Microsoft, 2016 Trends in Cybersecurity: A quick Guide to the Most Important Insights in Security, 2016. [Online]. Available: https://info.microsoft.com/rs/157-GQE-382/images/EN-MSFT-SCRTY-CNTNT-eBook-cybersecurity.pdf.
25. Chen, Hsinchun, Roger HL Chiang, and Veda C. Storey. "Business intelligence and analytics: From big data to big impact." MIS quarterly 36, no. 4 (2012): 1165–1188.
26. Durumeric, Zakir, James Kasten, David Adrian, J. Alex Halderman, Michael Bailey, Frank Li, Nicolas Weaver et al. "The matter of heartbleed." In Proceedings of the 2014 Conference on Internet Measurement Conference, pp. 475–488. ACM, 2014.
27. Mahmood, Tariq, and Uzma Afzal. "Security analytics: Big data analytics for cybersecurity: A review of trends, techniques and tools." In Information assurance (ncia), 2013 2nd national conference on, pp. 129–134. IEEE, 2013.
28. Talia, Domenico. "Toward cloud-based big-data analytics." IEEE Computer Science (2013): 98–101.
29. C. Pettey and R. Van der Meulen, Gartner Reveals Top Predictions for IT Organizations and Users for 2012 and Beyond, 01-Dec-2011. [Online]. Available: http://www.gartner.com/newsroom/id/1862714.[Accessed:01-Feb-2017].
30. Kambatla, Karthik, Giorgos Kollias, Vipin Kumar, and Ananth Grama. "Trends in big data analytics." Journal of Parallel and Distributed Computing 74, no. 7 (2014): 2561–2573.
31. Simmhan, Yogesh, Saima Aman, Alok Kumbhare, Rongyang Liu, Sam Stevens, Qunzhi Zhou, and Viktor Prasanna. "Cloud-based software platform for big data analytics in smart grids." Computing in Science & Engineering 15, no. 4 (2013): 38–47.
32. Cuzzocrea, Alfredo, Il-Yeol Song, and Karen C. Davis. "Analytics over large-scale multidimensional data: the big data revolution!." In Proceedings of the ACM 14th international workshop on Data Warehousing and OLAP, pp. 101–104. ACM, 2011.

33. Ericsson, Gran N. "Cyber security and power system communication essential parts of a smart grid infrastructure." IEEE Transactions on Power Delivery 25, no. 3 (2010): 1501–1507.
34. Khurana, Himanshu, Mark Hadley, Ning Lu, and Deborah A. Frincke. "Smart-grid security issues." IEEE Security & Privacy 8, no. 1 (2010).
35. Bejtlich, Richard. The practice of network security monitoring: understanding incident detection and response. No Starch Press, 2013.
36. Desai, Anish, Yuan Jiang, William Tarkington, and Jeff Oliveto. "Multi-level and multi-platform intrusion detection and response system." U.S. Patent Application 10/106,387, filed March 27, 2002.
37. Mell, Peter, and Tim Grance. "The NIST definition of cloud computing." (2011).
38. Burger, Eric W., Michael D. Goodman, Panos Kampanakis, and Kevin A. Zhu. "Taxonomy model for cyber threat intelligence information exchange technologies." In Proceedings of the 2014 ACM Workshop on Information Sharing & Collaborative Security, pp. 51–60. ACM, 2014.
39. Barnum, Sean. "Standardizing cyber threat intelligence information with the Structured Threat Information eXpression (STIX)." MITRE Corporation 11 (2012).
40. O'Toole Jr, James W. "Methods and apparatus for auditing and tracking changes to an existing configuration of a computerized device." U.S. Patent 7,024,548, issued April 4, 2006.
41. Gerace, Thomas A. "Method and apparatus for determining behavioral profile of a computer user." U.S. Patent 5,848,396, issued December 8, 1998.
42. Gu, Tao, Hung Keng Pung, and Da Qing Zhang. "Toward an OSGi-based infrastructure for context-aware applications." IEEE Pervasive Computing 3, no. 4 (2004): 66–74.
43. Anderson, Douglas D., Mary E. Anderson, Carol Oman Urban, and Richard H. Urban. "Debit card fraud detection and control system." U.S. Patent 5,884,289, issued March 16, 1999.
44. Camhi, Elie. "System for the security and auditing of persons and property." U.S. Patent 5,825,283, issued October 20, 1998.
45. L. Widmer, The 10 Most Expensive Data Breaches | Charles Leach, 23-Jun-2015. [Online]. Available: http://leachagency.com/the-10-most-expensive-data-breaches/.
46. SINET Announces 16 Most Innovative Cybersecurity Technologies of 2016 | Business Wire, 19-Sep-2016. [Online]. Available: http://www.businesswire.com/news/home/20160919006353/en/SINET-Announces-16-Innovative-Cybersecurity-Technologies-2016.
47. C. Pettey and R. Van der Meulen, Gartner Reveals Top Predictions for IT Organizations and Users for 2012 and Beyond, 01-Dec-2011. [Online]. Available: http://www.gartner.com/newsroom/id/1862714.
48. C. Heinl and E. EG Tan, Cybersecurity: Emerging Issues, Trends, Technologies and Threats in 2015 and Beyond. [Online]. Available: https://www.rsis.edu.sg/wp-content/uploads/2016/04/RSIS$_$Cybersecurity$_$EITTT2015.pdf.
49. Kavitha, T., and D. Sridharan. "Security vulnerabilities in wireless sensor networks: A survey." Journal of information Assurance and Security 5, no. 1 (2010): 31–44.
50. B. Donohue, Hot Technologies in Cyber Security, Cyber Degrees, 03-Dec-2014.
51. Jeong, Jongil, Dongkyoo Shin, Dongil Shin, and Kiyoung Moon. "Java-based single sign-on library supporting SAML (Security Assertion Markup Language) for distributed Web services." In Asia-Pacific Web Conference, pp. 891–894. Springer Berlin Heidelberg, 2004.
52. Gro, Thomas. "Security analysis of the SAML single sign-on browser/artifact profile." In Computer Security Applications Conference, 2003. Proceedings. 19th Annual, pp. 298–307. IEEE, 2003.
53. Rees, Loren Paul, Jason K. Deane, Terry R. Rakes, and Wade H. Baker. "Decision support for cybersecurity risk planning." Decision Support Systems 51, no. 3 (2011): 493–505.
54. T. Reuille, OpenGraphiti: Data Visualization Framework, 05-Aug-2014. [Online]. Available: http://www.opengraphiti.com/.
55. McKenna, S., Staheli, D., Fulcher, C. and Meyer, M. (2016), BubbleNet: A Cyber Security Dashboard for Visualizing Patterns. Computer Graphics Forum, 35: 281–290. doi:10.1111/cgf.12904

56. Linkurious, Linkurious - Linkurious - Understand the connections in your data, 2016. [Online]. Available: https://linkurio.us/.
57. T. Software, Business Intelligence and Analytics I Tableau Software, 2017. [Online]. Available: https://www.tableau.com/.
58. P. Corporation, Data Integration, Business Analytics and Big Data I Pentaho, 2017. [Online]. Available: http://www.pentaho.com/.
59. Norse Attack Map, 2017. [Online]. Available: http://map.norsecorp.com/$#$/.
60. Kaspersky Cyberthreat real-time map, 2017. [Online]. Available: https://cybermap.kaspersky.com/.
61. FireEye Cyber Threat Map, 2017. [Online]. Available: https://www.fireeye.com/cyber-map/threat-map.html.
62. Cyber Threat Map, FireEye, 2017. [Online]. Available: https://www.fireeye.com/cyber-map/threat-map.html.
63. L. SAS, data visualization Archives, Linkurious - Understand the connections in your data., 2015.
64. Interpol, Cybercrime / Cybercrime / Crime areas / Internet / Home - INTERPOL, Cybercrime, 2017. [Online]. Available: https://www.interpol.int/Crime-areas/Cybercrime/Cybercrime.
65. Nakamoto, Satoshi. "Bitcoin: A peer-to-peer electronic cash system." (2008): 28.
66. Barber, Simon, Xavier Boyen, Elaine Shi, and Ersin Uzun. "Bitter to better: how to make bitcoin a better currency." In International Conference on Financial Cryptography and Data Security, pp. 399–414. Springer Berlin Heidelberg, 2012.
67. Swan, Melanie. Blockchain: Blueprint for a new economy. " O'Reilly Media, Inc.", 2015.
68. IsecT Ltd, ISO/IEC 27001 certification standard, 2016. [Online]. Available: http://www.iso27001security.com/html/27001.html.
69. ISO, ISO/IEC 27001 - Information security management, ISO, 01-Feb-2015. [Online]. Available: http://www.iso.org/iso/iso27001.
70. IsecT Ltd, ISO/IEC 27032 cybersecurity guideline, 2016. [Online]. Available: http://iso27001security.com/html/27032.html.
71. Ware, Colin. Information visualization: perception for design. Elsevier, 2012.
72. Ramanauskait, Simona, Dmitrij Olifer, Nikolaj Goranin, Antanas enys, and Lukas Radvilavi-ius. "Visualization of mapped security standards for analysis and use optimisation." Int. J. Comput. Theor. Eng 6, no. 5 (2014): 372–376.
73. Deep Node, Inc, Why Deep Node?, Deep Node, Inc., 2016. [Online]. Available: http://www.deepnode.com/why-deep-node/.
74. Deep Node, Inc, The Concept Deep Node, Inc., 2016. [Online]. Available: http://www.deepnode.com/the-concept/.

Jeffery Garae is a PhD research student with the Cyber Security Researchers of Waikato (CROW). As a PhD candidate, his research focus is on security visualization for mobile platforms and user-centric visualization techniques and methodologies. He is also interested in data provenance, threat intelligence, attribution, digital forensics, post-data analytics and cyber security situation awareness. He values the importance of security in ICT. He is the first recipient of to the University of Waikato's Master of Cyber Security (MCS) program in 2014. He is currently the Doctoral Assistant for the Cyber Security course at the University of Waikato. In the ICT and Security industry, he has a great number of years experience with Systems and Networks. As part of his voluntary contribution to the Pacific Island countries, he serves as a security advisor and an advocate to Cyber Security Situation Awareness.

Ryan K.L. Ko is Head of the Cyber Security Researchers of Waikato (CROW) and Senior Lecturer with the University of Waikato. With CROW, he established NZ's first cyber security lab and graduate research programme in 2012 and 2013 respectively. He is principal investigator of MBIE-funded (NZ$12.23million) STRATUS project. Ko co-established the NZ Cyber Security Challenge since 2014. His research focuses on returning data control to users, and challenges in cloud computing security and privacy, data provenance, and homomorphic encryption. He is also interested in attribution and vulnerability detection, focusing on ransomware propagation. With 50 publications including 3 international patents, he serves on 6 journal editorial boards, and as series editor for Elsevier's security books. He also serves as the editor of ISO/IEC 21878 Security guidelines in design and implementation of virtualized servers. A Fellow of Cloud Security Alliance (CSA), he is a co-creator of the (ISC)2 CCSP certification—the gold-standard international cloud security professional certification. Prior to academia, he was a HP Labs lead computer scientist leading innovations in HP global security products. He is technical adviser for the Ministry of Justice's Harmful Digital Communications Act, NZ Minister of Communications Cyber Skills Taskforce, LIC, CSA and Interpol.

Printed in the United States
By Bookmasters